Superior Beings
If They Exist,
How Would We Know?

The central question posed in this book is: If there existed a superior being who possessed the supernatural qualities of omniscience, omnipotence, immortality, and incomprehensibility, how would he/she act differently from us? The mathematical theory of games is used to define each of these qualities, and different assumptions about the rules of play in several theological games that might be played between ordinary human beings and superior beings like God are posited. Implications of these definitions and assumptions are developed and used to explore such questions as: Are God's superior powers compatible with human free will? Can they be reconciled with the problem of evil in the world? In what situation is God's existence "decidable" in game-like relationships He might have with us?

By endowing omniscience/onmipotence/immortality/imcomprehensibility with unambiguous meanings, the author shows how game theory can help breathe life into questions that have been dismissed too quickly simply because they are metaphysical—outside the world of experience. Thereby he clarifies the structure of our thought about an ultimate reality, whether or not it is viewed as religious.

STEVEN J. BRAMS

Superior Beings
If They Exist,
How Would We Know?

Game-Theoretic Implications
of Omniscience,
Omnipotence, Immortality,
and Incomprehensibility

With 32 Illustrations

Springer-Verlag
New York Berlin Heidelberg Tokyo

Steven J. Brams
Department of Politics
New York University
New York, NY 10003
U.S.A.

AMS Subject Classifications: 00A69, 90D05

Library of Congress Cataloging in Publication Data
Brams, Steven J.
  Superior beings.
  Bibliography: p.
  1. God—Attributes. 2. Game theory. I. Title.
BT130.B69   1983   231'.4   83-10530

Typeset by University Graphics, Inc., Atlantic Highlands, NJ.
Printed and bound by R. R. Donnelley & Sons, Harrisonburg, VA.
Printed in the United States of America.

9 8 7 6 5 4 3 2 1

ISBN 0-387-91223-1 Springer-Verlag New York Berlin Heidelberg Tokyo (hardcover)
ISBN 0-387-90877-3 Springer-Verlag New York Berlin Heidelberg Tokyo (softcover)
ISBN 3-540-90877-3 Springer-Verlag Berlin Heidelberg New York Tokyo (softcover)

To Wayne A. Kimmel

Preface

The central question I pose in this book is: If there existed a superior being who possessed the supernatural qualities of omniscience, omnipotence, immortality, and incomprehensibility, how would he/she act differently from us, and would these differences be knowable? (Because God, the superior being in the Judeo–Christian tradition, is generally described as a male, I shall henceforth use the masculine pronoun form for convenience, but I intend no invidious gender distinctions, whether applied to supernatural or natural beings.) Theologians, philosophers of religion, and erudite scholars in other disciplines have addressed this and related questions before, but their answers, generally speaking, have not been informed by any systematic or rigorous theory.

I believe the mathematical theory of games, which has little to do with the frivolity and playfulness we normally associate with games, provides a powerful tool for clarifying the key theological concepts in my central question and drawing out their implications in games played between human and superior beings. I am fully aware that not everybody will agree that omniscience, omnipotence, immortality, and incomprehensibility are what I say they are, but I invite them to propose their own definitions and derive their own conclusions with the aid of the theory.

By endowing these protean concepts with unambiguous meanings, I will try to show how game theory can breathe life into questions that have been dismissed too quickly simply because they are metaphysical—outside the world of experience. In much of traditional philosophy, in my opinion, the abstract characteristics of God have been formulated in such nebulous or all-encompassing terms that they have been drained of significant content and have thereby suffered a rather pallid intellectual existence. Irredeemably metaphysical these characteristics may be, but this does not mean they are beyond the pale of analysis that helps to clarify the structure of our thought about an ultimate reality, whether or not it is viewed as religious, and thereby enhances our understanding of our place in the world.

Admittedly, ordinary humans "playing games" with some supernatural figure or force like God may sound fantastic if not absurd. But, I would argue, virtually all relationships we have with others—cooperative or antagonistic, shallow or deep, earthly or transcendent—can be characterized as games in a formal sense. Since theology is sometimes defined as the systematic study of our relationship with God, it seems to me entirely appropriate to use game theory to try to shed light on this spiritual, perhaps mystical, relationship, mundane as our conceptual apparatus might be. Indeed, I believe the austerity and parsimony of game theory facilitates both abstracting important aspects of this relationship and placing them within a unified framework.

Let me make clear that this kind of "mathematical" theology is not meant to reduce great religions and profound existential questions to mere numbers. Indeed, hardly any numbers are used in a quantitative sense in this book, though they are used to specify mathematical relationships, such as preference rankings in a game. Nonetheless, proponents of fideism, like Søren Kierkegaard, would undoubtedly be appalled by the application of any kind of objective reasoning to an understanding of religious faith, but most theologians are not so disparaging, and some (such as Paul Tillich, quoted at the end of this Preface) have welcomed it.

The heart of this study is an inquiry into (i) the meaning of superiority in games and (ii) the effects that different superior abilities have on the outcomes of such games. The effects are manifold, but in the end I stress the difficulties connected with ascertaining "decidability," by which I mean the ability of a human being who is in a game-playing relationship with a superior being to decide whether that being is indeed superior.

In a significant number of games, it turns out, this determination cannot be made because the superior being could not improve upon the (inferior) outcome that an ordinary player in his position could also achieve. Such undecidability suggests a new and strange kind of logic and may underlie agnosticism. In fact, the awesome mystery that undecidability engenders is reinforced by the fact that it may be rational in certain games for the superior being to obfuscate his choices by acting arbitrarily.

The main theological issue raised by arbitrariness and the attendant incomprehensibility it induces is whether such behavior is consistent with the actions of a benevolent and righteous God. This has been called the "problem of evil" by theologians and philosophers. I also probe other issues related to religious faith, including the degree to which a superior being's power may intrude on a human being's free will.

The sources of my ideas are diverse: from political science, a long-standing interest in the definition, measurement, and exercise of power; from mathematics, more than a decade's involvement with game theory and a particular fascination with its applications; from religion, an interest in the Bible stemming from a previous book in which I applied game theory to the exegesis of stories of conflict and intrigue in the Old Testament; and from philosophy, an appreciation that certain abstract and general questions are worth asking, even if they cannot be answered scientifically.

This last point about the proper role of science in analytic-philosophical work of this sort merits brief attention. Although mathematics is commonly identified with the sciences, this book is emphatically not a work of science. Not only is there no experimentation or other empirical testing of the propositions developed herein, but there is no suggestion that there ever could be, except by extrapolation to certain real-world situations, such as ascertaining the impact of certain kinds of secular power in a political conflict.

In a theological context, on the other hand, I cannot even conceive of God or other supernatural phenomena as being scientifically testable. Supernatural, by definition, means above or beyond the natural and therefore not susceptible to observation and measurement, which are ineluctable hallmarks of the sciences.

These caveats notwithstanding, I believe philosophical inquiry and theoretical analysis can help us explicate our inchoate thoughts about superior beings, their presumed uncanny attributes, the possible impact of these attributes on game outcomes, and what, finally, one might say about the decidability of superior beings from the nature of their game playing and the outcomes they induce. If this analysis does not lead to testable hypotheses, it will, I hope, provide a new way of looking at the metaphysical world that stimulates other thought experiments, grounded in logico-deductive analysis, in the ethereal realm. The heuristic value of a modern theory that gives a new slant to age-old questions should not be underestimated.

The game theory I use might appear arcane, but it is really quite elementary and, I believe, should be generally accessible to the nonmathematical reader who seriously wishes to follow the theoretical exposition. To facilitate this exposition, I have used a number of descriptive aids to highlight the key elements in the many figures in this book. In addition, I provide a Glossary of the more technical terms used at the end of the book.

I take leave of several key assumptions in classical game theory fairly quickly, so those familiar with this theory will have some surprises in store for them. I am not trying to be obtuse in these departures but rather am attempting to make the theory the handmaiden of the substance, not vice versa. At the same time, this focus, in my view, enriches the theory, too. As Paul Tillich argued in *Dynamics of Faith*, just as "reason is the precondition of faith," it "can be fulfilled only if it is driven beyond the limits of its own finitude, and experiences the presence of the ultimate. . . ."

New York   STEVEN J. BRAMS
April 1983

Acknowledgments

I have accumulated many intellectual debts and am happy to acknowledge them. First, seven people—Eva Brams, Marek P. Hessel, D. Marc Kilgour, Douglas Muzzio, Barry O'Neill, Laura J. Scalia, and Frank C. Zagare—read a first draft of this book, or major parts of it, from very different intellectual perspectives. They saved me from making several errors and called my attention to numerous ambiguities that I have tried to clarify. I am deeply appreciative even though I did not follow all their suggestions. John McDonald offered sage advice on the presentation of ideas at the beginning, and I thank him not only for this advice but also for his splendid encouragement of unorthodox applications of game theory over the years. Nancy Fernandez expertly typed the manuscript and also offered valuable comments, as did James D. Carse on the introductory material.

I am grateful to Marek P. Hessel and Donald Wittman for permitting me to use material from articles we co-authored (cited later) for purposes, I presume, they did not intend. Wayne A. Kimmel, to whom I dedicate this book, has not made such a direct contribution to this particular work, but our discussions on both large and small questions over twenty years have been a tremendous stimulation and delight to me. My wife Eva, and our children Julie and Michael, also provided much stimulation (and occasional provocation), which might well have given form and shape to some of my wilder pronouncements.

Contents

SIX Immortality and Incomprehensibility 123

SEVEN Superior Beings: They May Be Undecidable 157

List of Figures

Truly Thou are a hidden God.

Blaise Pascal, *Pensées*, 1670

No more *deus absconditus!*
Come out, come out, wherever you are, the game's over.

Walker Percy, *The Second Coming*, 1980

The contemplation of God's nature . . . will guide me through the tumult of life.

Albert Einstein, Letter To Johannes Stark, 1908

It is now acceptable for God to be seen . . . involved in the natural world of change; and at least in the view of some, not without dependency on man.

Richard Schlegal, "The Return of Man in Quantum Physics," in *The Sciences and Theology in the Twentieth Century*, 1981

What use, after all, is man, if not to teach God His lessons?

Peter Shaffer, *Amadeus*, 1980

God is something less than absolutely omnipotent. He is actually engaged in a conflict with his creature, in which he may very well lose the game. . . . Can God play a significant game with his own creature?

Norbert Wiener, *God & Golem, Inc.*, 1964

I [God] don't know how I do it. . . . Omniscience gives me eyestrain. . . . And omnipotence—*that* takes it out of you.

Stanley Elkin, *The Living End*, 1979

ONE

Introduction

Since the analysis in this book would seem far removed from work in any of the standard disciplines, I would like, in this introduction, to suggest some linkages that might not be apparent at first glance. This not only serves the purpose of establishing ties to relevant research in different fields but also helps to embed the present work in a research tradition, somewhat in eclipse, that I think still deserves to be part of the intellectual landscape.

This tradition weds broad philosophical questions to rigorous analytic methods, primarily developed in the sciences, for elucidating them. I have already disavowed the label of "science" being applied to this kind of study, because empirical investigation of the theological and philosophical questions I address seems out of the question.[1] In a secular context, however, predictability,

---

1. In 1940 Albert Einstein said, "Science without religion is lame, religion without science is blind" [Albert Einstein, *Out of My Later Years* (New York: Philosophical Library, 1950), p. 28], though the context of this statement suggests that by "religion" Einstein had in mind an abiding faith in the order of the universe rather than a personal God, or the activities associated with organized religions today. This viewpoint is supported by his statement, "When I am evaluating a theory, I ask myself, if I were God, would I have made the universe in that way?" [*Some Strangeness in the Proportion: A Centennial Symposium to Celebrate the Achievements of Albert Einstein,* ed. Harry Woolf (Reading, MA: Addison-Wesley, 1980), p. 476.]

power, and truth are amenable to empirical analysis in real-life games, and I hope my theoretical analysis of these and other concepts and their game-theoretic implications stimulates research in applied fields.

But this book is first and foremost a philosophical investigation of characteristics of superior beings and their possible impact on games played with human beings. It differs from a conventional philosophical inquiry in its relentless use of a mathematical theory that I have tried to adapt in order to (i) facilitate the conceptualization of superiority, (ii) deduce the consequences superiority has on game outcomes, and (iii) explore theological issues raised by a disciplined inquiry of this sort.

By contrast, the primary tool of inquiry in philosophy is logic—in its various manifestations—which, while "mathematical" in a certain sense, is not really a theory about anything except what constitute correct inferences and valid arguments. This is a fundamental question in all disciplines, to be sure, but it is best asked of a theory that provides one with a specific orientation to a substantive problem and is in need of incisive scrutiny and critical assessment.

Game theory offers a way of analyzing situations of conflict and cooperation in which the actors, or players, are assumed to make rational choices. To act *rationally*, in simplest terms, means to choose strategies that lead to better rather than worse outcomes in a game. This choice will depend not only on one's own goals, which I try to justify in most games discussed in this book, but also on the goals and rational choices of other players. In light of the contingent nature of rational choices in a game, game theory can properly be viewed as a theory for making optimal strategic calculations that take into account these contingencies, uncertainties in the environment, and so on.

A *game* is sometimes defined as the sum-total of the rules that describe it. For the most part, I shall define a game by an "outcome matrix" or "game tree" and certain "rules of play," which will be given precise meanings later. Because the outcome of a game in this form depends on the rational strategic choices of *all* players, it is fair to say that such games describe truly interdependent decision situations.

No technical knowledge of game theory is assumed in this book, which I have tried to write for the lay reader who may, nevertheless, have difficulty following all the steps in an argument. I would urge this reader not to get bogged down at troublesome points but to go on and first see where the analysis leads before turning back to try to understand all the details. The stumbling blocks will often disappear, and the pieces fall into place, when the results of the deductions and calculations come into view.

Later I shall say much more about the game-theoretic calculations used here, which by no means mirror all the major branches of the theory. In fact, I utilize only the so-called noncooperative theory, in which communication between players is allowed but binding agreements and enforceable contracts are not permitted. Moreover, I analyze only two-person games, and sometimes one-person games against nature, so the cooperative theory dealing with coalitions, and what they can ensure for their members, is omitted.

In the two-person setting, however, I extend the classical theory to allow for dynamic sequential play between an ordinary being and a superior one, who possesses the various supernatural qualities I shall describe shortly. How these qualities of a superior being affect the sequences of moves that will by played, and the outcomes that will be implemented, will be systematically examined in the case of each of these qualities, and then all together in Chapter 7. In the Appendix, I summarize the main technical results of this book for the 57 distinct 2×2 ordinal games of conflict in which the two players, each with two strategies, can rank the four outcomes from best to worst but disagree on a most-preferred outcome.

How does one define "supernatural" and still stay close to concepts of the natural world we know best? Research in the field of artificial intelligence provides some clues. Analysts in this field have struggled for over a generation with the question of how one can distinguish human intelligence from nonhuman, or artificial, intelligence.

Put in practical terms, can a computer think? Can it have consciousness or self-awareness? Can it experience human feelings? If not now, are these theoretical possibilities or impossibilities?

Needless to say, there is no consensus on answers to these questions. In fact, a great deal of controversy swirls around them and the proper definition of terms like "thinking" and "consciousness."

The most famous, and I believe still the most insightful, attack on this problem was made by A. M. Turing.[2] He suggested a test whereby an interrogator, by asking questions of a man and woman, would try to determine which was the man and which was the woman.

The man, he assumed, would try to fool the interrogator, perhaps by pretending to be the woman, while the woman would try to help the interrogator, presumably be being truthful. In answering the questions, both would be separated from the interrogator, and identifying characteristics like voice would be rendered indistinguishable by use of, say, a teleprinter in communication with the interrogator.

Now if the man were replaced by a machine/computer designed or programmed to try to deceive the interrogator—just like the real man—would the interrogator do any better in distinguishing the woman from her (artificial) imitator? If the answer were no, Turing argued, it would be fair to say that the artificial intelligence was on a par with the human intelligence in its deception (thinking?) ability, for the interrogator would not be able to discriminate between the two.

I relate all this to make the point that I believe the question of the decidability of a superior being is a similar one, though on a different level. In a way, Turing's imitation game is turned on its head, because one is not interested in whether one can *mimic* human intelligence but whether one can *distinguish* superior from human intelligence when the superior and the human players are pitted against each other in a game.[3]

This higher-level game introduces new difficulties, however.

---

2. A. M. Turing, Computing machinery and human intelligence, *Mind* 59, 236 (October 1950), 433–460. For collections of recent views and approaches to this problem, see *Mind Design: Philosophy, Psychology, Artificial Intelligence*, ed. John Haugeland (Cambridge, MA: MIT Press, 1981); and *The Mind's I: Fantasies and Reflections on Self and Soul*, ed. Douglas R. Hofstadter and Daniel C. Dennett (New York: Basic, 1981). For the apostate view that ancient man's relationship with God (or gods) might have involved direct communication with Him (or them), physiologically based, within the brain itself, see Julian Jaynes, *The Origin of Consciousness in the Breakdown of the Bicameral Mind* (Boston: Houghton Mifflin, 1976).

3. The problem may be compounded if the superior being is a sentient person. Consider, for example, John von Neumann (1903–1957), the great twentieth-century mathematician and cofounder, with economist Oskar Morgenstern (1902–1977), of game theory: "The story used to be told about him [von Neumann] in Princeton that while he was indeed a demi-god he had made a detailed study of humans and could imitate them perfectly." [Herman Goldstine, *The Computer from Pascal to von Neumann* (Princeton, NJ: Princeton University Press, 1972), p. 176.]

Although human intelligence and creativity are hard to pin down conceptually—much less measure—specifying what a supernatural intelligence is seems an even more challenging and intractable problem for a number of reasons.

First, there are no examples to observe. True, an extraterrestrial intelligence that presented itself might give us an empirical case for study, but would we have the wit to understand the signals it emits, which may or may not be intended to inform us truthfully? Furthermore, there is the more basic epistemological question: How would we even know that it is a superior intelligence if such a thing had never been observed before? In addition, our understanding is limited by our own ordinary intelligence.

These may not be insuperable difficulties, but I would prefer to set them aside for now and look instead at what religion has to say about some of these matters. Of course, there are no single answers, and, even worse, orthodoxy in one faith is heresy in another. Nevertheless, there is common ground, centered on the ineffable, which I refer to as the "mystery of it all," or, for future reference, the Great Mystery.

Now different religions divine the Great Mystery differently, but none expunges it completely. Indeed, I contend that the shared feature of all great world religions is a belief that no kind of inquiry, scientific or otherwise, can clear the air of the spiritual element hovering over—or immanent in—our lives.

From a religious viewpoint, this is the quintessential element in our existence, and we should celebrate it, not try to snuff it out of our lives or dismiss it as purely metaphysical and, therefore, unworthy of study. The spiritual cannot be purged from our lives anyway, believers would say. Besides, we may gain inspiration if not insight from the sacred texts which glorify the Great Mystery.

The inspirational provenance of many of my own ideas on superior beings is the God of the Hebrew Bible, or Old Testament. This Bible, of course, is used in some form, though not always for the same purposes, by Judaism, Christianity, and Islam. As I have argued elsewhere,[4] God is a complex and fascinating figure and, though perhaps a figment, Someone who appears in the biblical narratives as a superlative strategist whose rational actions are driven by very humanlike calculations as well as emotions.

I believe there is no more fertile source on the characteristics of superior beings than the biblical God. Throughout this book, I

---

4. Steven J. Brams, *Biblical Games: A Strategic Analysis of Stories in the Old Testament* (Cambridge, MA: MIT Press, 1980).

use Him and other biblical characters, and their actions in game-like situations, to justify various strategic assumptions and to lend plausibility to definitions of superior qualities that I offer. The concreteness of the biblical stories provides, in addition, a palpable counterpoint to the abstract and perhaps recondite theory I shall develop and pursue later.

This is not to say that the Bible offers the final word on philosophical and theological matters in the modern world. Nonetheless, it seems to me to be a logical place from which to start, and the clues it offers on God's preferences and powers do not seem to me contradicted by contemporary events.

This is a sharply different perspective from most contemporary work in the philosophy of religion with which I am familiar.[5] There is hardly any mention of religious sources, except in passing, and then not usually for the purpose of acquiring understanding from any kind of careful exegesis. When God is mentioned, which occurs more frequently, it is usually for the purpose of conjuring up philosophical conundrums about His powers (Can He create a stone so heavy He cannot lift it?), which often have no basis outside of the philosopher's imagination and the remote abstractions of theology.

---

5. See, for example, the articles in *The Logic of God: Theology and Verification*, ed. Malcolm L. Diamond and Thomas V. Litzenburg, Jr. (Indianapolis: Bobbs-Merrill, 1975); and *Contemporary Philosophy of Religion*, ed. Steven J. Cahn and David Shatz (New York: Oxford University Press, 1982). As an introduction to the field, I would commend Leszek Kolakowski, *Religion* (New York: Oxford University Press, 1982), which is both a slightly irreverent intellectual history and a provocative theoretical synthesis of "God, the Devil, Sin and other Worries of the so-called Philosophy of Religion." Kolakowski takes the position, with which I concur, that "God is not and cannot be an empirical hypothesis . . . if the word 'hypothesis' retains its usual sense" (p. 90). That is why "no one has ever heard of God's existence being discussed at conferences of physicists . . . as science offers no conceptual tools with which to tackle the problem" (p. 67).

My own view is that the conceptualization of a superior being is an entirely different matter from that of the empirical testing and verification of a superior being, such as an extraterrestrial intelligence. We can move ahead on the conceptual issues even if science holds out little hope at this time for corroborating our theoretical analysis. Just as theoretical analysis has established that there are certain kinds of problems a computer can never solve, I shall argue in the end that there are certain kinds of limits to our ability to ascertain the "decidability" of superior beings, as I conceptualize them, which I think is an important result even if it has no empirical interpretation that would allow it to be tested scientifically. Again, my purpose is to offer a philosophical perspective or point of view, not a scientific theory, my use of mathematics notwithstanding.

My approach is radically different. In my opinion, it is eminently well suited to characterizing our possible relationship with a personal God, which I shall discuss shortly. I suspect, however, that it will be criticized for its seeming sacrilegious reduction of superior beings to political-strategic actors, without goodness, grace, mercy, and other virtues we normally associate with God. In brief rebuttal, I would say that *Biblical Games* presents my case for God the game player, who, by and large, constrains but does not abrogate the ability of humans to make free choices. More to the point, though, there is nothing in game theory that says players must be odious, sinister, or vile creatures, or purely strategic animals, for that matter.

On the contrary, all preference configurations in which players can strictly rank outcomes from best to worst are possible in the games I analyze. In many, the superior and human players have symmetrical preferences, whereby their likes and dislikes are mirror images of each other in a sense to be made precise later; in a few, they have exactly the same preferences for each outcome. Although both players are assumed to be rational in the sense of desiring to bring about their best possible outcomes, they differ in the means at their disposal.

I develop and interpret a relatively small number of games in considerable detail in each of the chapters—and sometimes over several chapters—to illustrate the nature of rational belief and the different effects of omniscience, omnipotence, immortality, and incomprehensibility. Let me now indicate, in general terms, what I mean by these concepts.

Omniscience has the most straightforward definition—the ability of the superior player to predict the human player's strategy choice in a game—but it gives rise to a paradox when the human player, aware of the superior player's foreknowledge, exploits it. Yet the moves and countermoves I assume the players can make in sequential play allow the superior player to circumvent this paradox, as I shall show, though not a second omniscience paradox that omnipotence (described below) allows him to overcome.

Two concepts of *omnipotence* are proposed and their consequences developed: "moving power," which roughly means the ability to continue moving in a game when the other player must eventually stop; and "staying power," which refers to the ability to delay making a strategy choice until the other player has made his. Neither of these qualities makes the superior being the all-powerful figure "omnipotence" connotes. But, as I argue, making the superior being an absolute dictator would be inconsistent with the human player's having free will, and being able to make independent choices, in a game. Although the two different kinds of omnipotence enable the superior being to induce the same outcome in most games, in some games their effects are different. So are the effects of *immortality*, which I tie to "threat power," because it is rational to use such power only in repeated plays of a game. Precisely because he is immortal, it is rational for the superior player to threaten a mutually disadvantageous outcome in the single play of a game to deter untoward actions by the same or different mortal players in the future play of this or other games.

Incomprehensibility is defined in terms of an extension of so-called mixed strategies to nonconstant-sum games, in which it may be optimal to make certain random strategy choices. Such randomness by the superior player may be interpreted as arbitrary behavior, which I suggest offers a partial explanation of the alarming problem of evil and suffering in the world: even with worthy goals, such as trying to elicit the truth from the human player, it still may be rational for the superior player to act arbitrarily (randomly) on occasion.

Finally, I identify *undecidable games*, in which the superior player does worse than the ordinary player, his superiority notwithstanding. That arbitrariness may contribute to undecidability in other games exacerbates the problem of distinguishing the superior player by his strategy choices/moves and the outcome of a game.

An indistinguishable or arbitrary superior being poses, I think, a fundamental challenge to those for whom God's existence and goodness are self-evident. While God may well exist and be benevolent, there are, I believe, thorny problems in demonstrating this, even in principle, in games that may involve both Him and us as players (the western religious bias of this conceptualization is discussed below). Likewise, the same problems beset proving God's nonexistence. Together, they provide a heretofore unexplored argument for agnosticism, which seems to me particularly powerful because it says that the problem is theoretical: its solution does not simply depend on finding better evidence (whatever this may be), whereas, traditionally, agnosticism has represented a timid compromise between theism and atheism.

Because the power-based concepts, especially, may strike

one as too limited, naturalistic, and "political" to capture the full-blownness and augustness of omnipotence and immortality, the conclusion about the undecidability of superior beings in games may seem unwarranted. Short of making a superior being like God a tyrant or a dictator, however, which would contravene a human being's free will and sovereignty, it is unclear what else these divine attributes convey.

In story after story in the Bible, it is quite plain that man's choices are not forced upon him by God (more details will be given later), so I think it is proper not to assign a meaning to the concept of omnipotence, in particular, that would deprive human players of their freedom and autonomy. Although it might seem a contradiction in terms, omnipotence, I believe, must in a sense be partial. To make it too inclusive a power would render it cognitively empty: it would be the root force lurking behind everything—and hence would explain almost nothing—and would also contradict my assumption that human beings have free will and can make their own unfettered choices that affect the outcome of a game.

Even in the everyday world of politics, the concept of power has proved notoriously difficult for political scientists to reach any consensus on, both with respect to its conceptualization and its measurement. Surely, then, it is not surprising that omnipotence raises issues at least as formidable, though, in the absence of verifiable observations, there is no measurement problem.

More fundamentally, if one challenges the game-theoretic framework itself, especially the seemingly impious notion that a superior being such as God plays games with us, some of the difficulties I alluded to earlier may lose their force, if not vanish. But most western religions view human beings and God to be somehow connected, which rests on the notion that God can be conceived of in personal terms—is, in fact, a personal God in a one-to-one relationship with each of us.

The idea of a personal God—especially One who is a game player—is alien to most eastern religions. So is the notion that God is a lofty and majestic figure, Someone of overpowering grandeur and infinite wisdom and strength. I confess that the present analysis has a western religious bias, perhaps epitomized by the view of Martin Buber:

> The description of God as a Person is indispensable for everyone who like myself means by "God" not a principle . . . not an idea . . .

but who rather means by "God," as I do, him who—whatever else he may be—enters into a direct relation with us.... [6]

However, despite this focus on a personal God, in Chapter 2 I consider the possibility that God might be more accurately conceived of as a "state of nature"—not a player as such—with the choice of nature being neutral or indifferent and made according to some random device.

This randomness should not be confused with the random strategies that are discussed in Chapter 6, which are probabilistic choices, too, but optimal for an active player. Nonetheless, there is an arbitrariness that underlies the notion of random behavior, purposeful or not, which surely impedes a deterministic understanding of either a sublime God or a "natural order" of things. More specifically, it suggests that such understanding must be probabilistic, and hence at core somewhat mysterious, even when our efforts to unmask the Great Mystery are unflagging.

I began this chapter by saying that the tradition of attacking important questions in philosophy and theology with the analytic tools of the sciences—if not their verification procedures—seems to be on the wane. A contemporary Russian mathematician, I. R. Shafarevitch, has decried the failure of mathematics to address such questions, but he held out the hope that

... mathematics may serve now as a model for the solution of the main problem of our epoch: to reveal a supreme religious goal and to fathom the meaning of the spiritual activity of mankind.[7]

6. Martin Buber, *I and Thou*, 2nd Ed., translated by Ronald Gregor Smith (New York: Scribner's, 1958), Postscript, p. 135; see also Emil L. Fackenheim, An outline of modern Jewish theology, in *Faith and Reason: Essays in Judaism*, ed. Robert Gordis and Ruth B. Waxman (New York: KTAV, 1973), pp. 211–220; and Abraham Kaplan, The Jewish argument with God, *Commentary* 70 (October 1980), 43–47, for more on the theology of Judaism. Albert Einstein had this to say about a personal God: "The doctrine of a personal God interfering with natural events could never be refuted, in the real sense, by science, for this doctrine can always take refuge in those domains in which scientific knowledge has not yet been established." [Phillip Frank, *Einstein: His Life and Times* (New York: Knopf, 1947), p. 285.] It is not evident to me whether Einstein, in using the phrase "has not yet been established," was at all sanguine that science might eventually say more.

7. Quoted in Philip J. Davis and Reuben Hersh, *The Mathematical Experience* (Boston: Birkhäuser, 1980), p. 54. Another mathematician has echoed this sentiment in a recent thoughtful book on infinity and its far-reaching ramifications:

... It is important to realize that such traditional questions as "Can we know the Absolute?," "Is Reality One or Many?," or "What is Truth?" are real questions that can be investigated in an exact way. An unfortunate

Although I would be circumspect about applying the word "solution" to Shafarevitch's "main problem," I would hope that the interest in the intellectual tradition I spoke of earlier, spurred by new ideas and invigorated by new methods, will be renewed, and a kind of speculative analytical philosophy will be rejuvenated.

I know of no reasons, in principle, why some of the great religious-theological-philosophical questions of our age cannot be made more perspicuous, their analysis more coherent, and their implications better understood by the use of formal deductive methods appropriate to the problem at hand. The problems will vary, and so will the methods, but the marriage, if consummated, could have auspicious prospects. It will, I trust, not be dull.

---

effect of the early logical positivism was that for many years professional philosophers tended to dismiss the ultimate metaphysical questions as woolly at best and meaningless at worst. . . . These [big] questions, far from being meaningless, can lead to good and exciting mathematical philosophy of the highest order. [Rudy Rucker, *Infinity and the Mind: The Science and Philosophy of the Infinite* (Boston: Birkhäuser, 1982), p. 218.]

TWO

The Rationality of Belief in a Superior Being

2.1. Introduction

In this chapter I shall consider the question of whether it is rational to believe in a Superior Being (SB), who may be thought of as God, or some other religious figure, or a secular force. I shall not stress the religiosity of SB, but I shall allude to religious works, particularly the Bible, to try to ferret out and understand what the goals of SB might be in games I postulate he plays with Person (P), the human player who must decide whether or not to believe in SB's existence. A few technical terms will be introduced in this chapter, but all will be illustrated in the games and decisions that are analyzed.

The rationality of religious beliefs is a venerable question that has been the subject of an enormous body of literature, both classical and modern. The most famous rational argument for believing in God is that of Pascal, whose wager to justify such belief I shall describe in Section 2.5.

Pascal's approach is decision-theoretic—he does not assume we play games against God, in which God actively chooses strategies. Instead, he supposes that each person makes calculations of a particular kind about whether belief, in an uncertain world, is justified.

At the outset, I shall assume that SB is not indifferent but instead makes strategy choices in two different kinds of games, the first being represented by what I call the Revelation Game, the second by four separate but related Knowability Games. I shall show that there is a paradox associated with rational choices in the former game; one of the latter games, known in the game theory literature as Prisoners' Dilemma, is vulnerable to a similar paradox. The other Knowability Games suggest different answers to the question of whether it is rational for P to expect SB to be knowable, depending on P's preferences as well as SB's.

Next, in two separate decisions, I consider whether it is rational for P to search for indications of SB's existence, or be concerned about him, depending on whether SB is aware of, or cares about, P. The spirit of Pascal's wager is captured in these decisions, though the surety of betting on God's existence is shown to collapse when Pascal's special assumptions about P's payoffs are dropped. Specifically, P has no unequivocally best choice, based only on his assumed rankings of the outcomes, when he must make choices in the face of uncertainty.

What I do not address in this chapter are the so-called rational proofs of God's existence, as well as those of nonexistence. In a superb synthesis of the literature, beginning with Descartes, on the existence of God, Hans Küng argues that all such proofs are flawed,[1] and I share his view. In place of proof, Küng argues for a "rationally justified" faith, which of course does not have the force of a logical proof but is, as the subtitle of his book suggests, "an answer for today." This answer, according to Küng, is rooted to developing a fundamental "trust in reality."

The question of whether this answer is satisfactory as a religious response I leave to theologians and philosophers of religion. Instead, as I have indicated, I want to focus at the start on a related, if more subjective, question: Is it rational to *believe* in SB's (or God's) existence? (The rationality of theistic belief is separate from its truth—a belief need not be true or even verifiable to be

---

1. Hans Küng, *Does God Exist? An Answer for Today*, translated by Edward Quinn (New York: Doubleday, 1980). For arguments that God's existence is "probable" but not "indubitable," see Richard Swinburne, *The Existence of God* (Oxford: Clarendon, 1979); to me Swinburne's calculations smack of bogus quantification.

rational[2]—though I shall suggest a connection between the two below.) Küng treats this question as well in *Does God Exist?*, and also in a much shorter book, *Freud and the Problem of God.*[3] However, I propose to focus less on the psychological aspects of belief, related to satisfying certain emotional needs, and more on its rational bases, tied to the "evidence" at hand.

The evidence for believing or not believing in the existence of God depends, in part, on whether God reveals Himself and, if so, makes Himself knowable or intelligible. These choices are not unrelated: one's revelation may be dubious if what is revealed is unintelligible or incommunicable; intelligibility, on the other hand, may be suspect if it can be understood without resorting to some mystical appearance or experience, which is to some the essence of revelation.

In the Bible, curiously, the question of God's existence is almost never raised. When it is, as in Moses's confrontation with Pharaoh in the Book of Exodus, it is the tangible evidence of God's miraculous powers that settles the issue for Pharaoh. For many today, however, the evidence is not so compelling.

2.2. *The Revelation Game*[4]

The Revelation Game is a two-person game of partial conflict (to be explained shortly), in which each player has two strategy choices. As I indicated earlier, using a simple game to model the relationship P might have with SB, or God, drastically simplifies a deep and profound religious experience for many people. My aim, however, is not to describe this experience but to abstract from it a central theological question: Can belief in SB be conceptualized as a rational choice if, by SB, one means an active entity who is capable of making choices? I shall approach this question by describing a game that assumes particular goals of P and SB, and then show that play of this game leads to a paradoxical outcome

2. George I. Mavrodes, Rationality and religious belief—a perverse question, in *Rationality and Religious Belief,* ed. C. F. Delaney (Notre Dame, IN: University of Notre Dame Press, 1979), p. 31.

3. Hans Küng, *Freud and the Problem of God,* translated by Edward Quinn (New Haven, CT: Yale University Press, 1979). For a collection of other views on this question, both ancient and modern, see *Rationality and Religious Belief.*

4. This and the next section are based largely on Steven J. Brams, Belief in God: a game-theoretic paradox, *Int. J. Philos. Religion* 13, 3 (1982), 121–129.

for both players. I shall also comment on how choices in the Revelation Game might bear on our knowledge of God's existence and, in addition, suggest some connections between the assumptions of the game, particularly concerning God's preferences, and the Bible.

In the Revelation Game, assume that SB has two *strategies*, which are the possible courses of action from which he chooses:[5]

1. Reveal himself (establish his existence)
2. Don't reveal himself (don't establish his existence)

(The parenthetic interpretations, concerned with SB's establishing his existence, will be discussed later.) Similarly, P has two strategy choices:

1. Believe in SB's existence
2. Don't believe in SB's existence

The intersections of these strategies define the four outcomes of the Revelation Game, shown in Fig. 2.1.

5. In more complex situations, illustrated by the Revised Punishment Game in Section 6.4 (Fig. 6.4), strategies are complete plans that take into account all possible choices that other players can make.

Figure 2.1   *Outcome Matrix of Revelation Game*

		P	
		Believe in SB's existence	**Don't believe in SB's existence**
	Reveal himself (establish his existence)	P faithful with evidence: belief in existence confirmed (3,4)	P unfaithful despite evidence: nonbelief in existence unconfirmed (1,1)
SB	**Don't reveal himself (don't establish his existence)**	P faithful without evidence: belief in existence unconfirmed (4,2)	P unfaithful without evidence: nonbelief in existence confirmed (2,3) ←Dominant strategy

Key: $(x,y) = $ (SB,P)

  4 = best; 3 = next best; 2 = next worst; 1 = worst
  Circled outcome rational (Nash equilibrium)

In a game in normal or matrix form, it is customary to represent an outcome by an ordered pair of numbers, (x,y), with the first number x being the preference of the row player (SB), the second number y the preference of the column player (P). *(Preference* simply means a ranking of the outcomes by a player.) I assume "4" represents the best outcome for a player, "3" the next best, "2" the next worst, and "1" the worst. Thus, the higher the number, the better the outcome, though I assume that no numerical value or cardinal utility is associated with each outcome. Because the numbers represent only ordinal preferences, one cannot say how much more a player prefers, say, the outcome he ranks 3 to the outcome he ranks 2.[6]

The ranks in the outcome matrix of Fig. 2.1 are based on the following assumptions about the goals of the two players, which I shall discuss in some detail shortly:

SB: (1) Primary goal—wants P to believe in his existence
    (2) Secondary goal—prefers not to reveal himself
P: (1) Primary goal—wants belief (or nonbelief) in SB's existence confirmed by evidence (or lack thereof)
    (2) Secondary goal—prefers to believe in SB's existence

The primary and secondary goals of each player *completely* specify their ordering of outcomes from best to worst: yes/no answers for each player and each goal automatically rank the four cells of a 2 × 2 game (a tertiary goal would rank the eight cells of a 2 × 2 × 2 game). This is an example of a lexicographic decision rule, whereby outcomes are first ordered on the basis of a most important criterion, then a next most important criterion, and so forth.[7]

In SB's case, the primary goal establishes that he prefers outcomes in the first column of the outcome matrix (believe in SB's existence) to outcomes in the second column (don't believe in SB's existence); between the two outcomes in each column, the secondary goal establishes that SB prefers not to reveal himself over revealing himself. In P's case, the primary goal says that he prefers to have his belief or nonbelief confirmed (main-diagonal out-

6. Note, though, that because the best outcome for one player is not worst for the other, etc., the game is not one of total conflict. (For an example of such a game, see Section 2.4.) Rather, since both players can, comparatively speaking, "win" [at (3,4)] or "lose" [at (1,1) or (2,3)] simultaneously, the game is one of *partial conflict.*

7. See Peter C. Fishburn, Lexicographic orders, utilities and decision rules: a survey, *Management Sci.* 20, 11 (July 1974), 1442–1471.

comes) to being unconfirmed (off-diagonal outcomes); between the pairs of main-diagonal and off-diagonal outcomes, the secondary goal says that P prefers to believe, rather than not believe, in SB's existence.

In the contemporary world, I would argue, evidence—from one's experiences, observations, and reflections—accumulates that predisposes one to believe or not believe in the existence of God or some other supernatural force, or leaves the issue open; but how beliefs are formed is less clear.[8] There are also, of course, religious sources that condition one's thinking; some brief remarks on the Hebrew Bible may lend plausibility to my goal assumptions for P and SB.

Evidence that the God of the Old Testament wanted His supremacy acknowledged by both Israelites and non-Israelites is abundant in the Bible. Moreover, the biblical narratives make plain that He pursued this goal with a vengeance not only by severely punishing those who did not adhere to His commands and precepts but also by bestowing rewards on the faithful who demonstrated their unswerving belief through good deeds and sacrifices.[9]

Yet, beyond providing indirect evidence of His presence through displays of His might and miraculous powers, God had an overriding reason for not revealing Himself directly: it would have undermined any true test of a person's faith, which I assume to be belief in God not necessarily corroborated by direct evidence. Only to Moses did God corroborate His existence directly—"face-to-face" (Exod. 33:11; Num. 12:6–8; Deut. 34:10)—but that Moses actually saw God firsthand is contradicted by the statement God made to Moses: "But," He said, "you cannot see My face, for man may not see Me and live" (Exod. 33:20).[10]

Since we cannot be tested if God's existence has already been confirmed by some unequivocal revelatory experience, I assume God most desires from us an expression of belief that relies only

---

8. For a developmental analysis of faith, see James W. Fowler, *Stages of Faith: The Psychology of Human Development and the Quest for Meaning* (San Francisco: Harper and Row, 1981). Different kinds of evidence, and the different kinds of rationality they give rise to, are discussed in Richard Swinburne, *Faith and Reason* (Oxford: Clarendon, 1981), Chaps. 2 and 3.

9. For the evidence in support of these contentions that goes beyond the cursory biblical citations I provide here and later, see Steven J. Brams, *Biblical Games: A Strategic Analysis of Stories in the Old Testament* (Cambridge, MA: MIT Press, 1980).

10. All biblical passages quoted in this book are from the following translations of the Jewish Publication Society, Philadelphia: *The Torah: The Five Books of Moses* (2nd Ed., 1967); *The Prophets* (1978); and *The Writings* (1982).

on faith (i.e., belief without direct evidence). Indeed, it is not unfair, in my opinion, to read the Bible as the almost obsessive testing of man by God to distinguish the faithful from those whose commitment to Him is lacking in zeal or persistence (remember that Job's faith faltered, but he never abandoned God).

This all-too-brief justification of SB's goals by way of the biblical God's will not be persuasive to those who regard the Bible as an unrealiable source at best, pure fantasy at worst. It is *not*, however, a nonbeliever whom I postulate as P in the Revelation Game. Instead, I assume that this is a person who takes the Bible (or other monotheistic religious works) seriously: the Bible or other works may well describe experiences that are outside or beyond the secular world, but P has yet to make up his mind about the existence of an "ultimate reality" embodied in some SB. While he entertains the possibility of SB's existence, and in fact would prefer confirmatory to nonconfirmatory evidence, *evidence is his key concern*. Moreover, P realizes that whether SB provides it or not will depend on what constitute "rational" choices for the players in the Revelation Game, to be discussed in Section 2.3.

Note that I do not assume that P can choose to believe in SB's *nonexistence*—only that he can choose not to believe in his existence, i.e., be agnostic. For P understands that SB might have good reasons for not revealing himself, so nonrevelation does not prove SB's nonexistence; its choice by SB in the Revelation Game only makes agnosticism for P preferable to believing. This second strategy choice of P can more completely be expressed as "don't believe in SB's existence *or nonexistence,*" which is to say, suspend judgment—a kind of commitment to remain noncommital.

2.3. *The Revelation Game Paradox*

To highlight the main problem, or paradox, that one's belief or nonbelief in SB's existence through possible revelation entails in a game-theoretic setting, assume in the subsequent discussion that there in fact exists a player, called SB, who can choose to reveal himself if he wishes to do so. If he has the goals/preferences I assumed earlier, he has a *dominant strategy* in the Revelation Game, namely not to reveal himself. That is, whatever P chooses, SB prefers his second strategy: if P believes in his existence (first column), SB prefers not to reveal himself because (4,2) is better for him than (3,4); if P does not believe in his existence (second column), SB still prefers not to reveal himself because (2,3) is better for him than (1,1). The fact that SB has an unconditionally best strategy choice—its superiority does not depend on what strategy P chooses—is, I think, a cogent reason for SB to choose it.

Note that P's own preferences do not give him a dominant strategy: belief is better for him if SB reveals himself, but nonbelief is better if SB does not, so P does not have an unconditionally best strategy choice. However, if the Revelation Game is one of complete information, as I assume it is, P knows SB's preferences as well as his own.

Knowing SB's preferences, P can infer that SB has a dominant strategy and will undoubtedly choose it. This being the case, since P prefers (2,3) to (4,2) in the second row of the outcome matrix, he would choose his own second strategy (second column) to ensure his preferred outcome, thereby making (2,3) the rational outcome of the game.

The foregoing reasoning is reinforced by the fact that (2,3) is the unique stable outcome, or *Nash equilibrium* in pure strategies, in this game: once chosen, neither player would have an incentive to depart unilaterally from this outcome.[11] For example, if SB switched to his first strategy, he would be worse off at (1,1); and if P switched to his first strategy, he would be worse off at (4,2). In the case of each of the other three outcomes, by comparison, at least one player has an incentive to depart unilaterally because he would do better if he did, rendering these outcomes unstable.

Yet (2,3) is only the next-worst outcome for SB and the next-best for P. Moreover, there exists another outcome, (3,4), that is preferred by *both* to (2,3), rendering the latter outcome a *Pareto-inferior* one [as is (1,1)]—that is, inferior for both players to some other outcome.[12] Thus, not only is it rational for SB not to reveal himself and for P not to believe in his existence—a problem in

11. See John Nash, Non-cooperative games, *Ann. Math.* 54 (1951), 286–295; and, more generally, Steven J. Brams, *Game Theory and Politics* (New York: Free Press, 1975), on this and related game-theoretic concepts. The classic work on game theory is John von Neumann and Oskar Morgenstern, *Theory of Games and Economic Behavior*, 3rd ed. (Princeton, NJ: Princeton University Press, 1953); the first edition of this book was published in 1944.

12. If not all preferences are strict (i.e., if players are indifferent about some outcomes), then a Pareto-inferior outcome is one that is worse for at least one player, and not better for the other player(s), than some other outcome. For example, if (2,3) in Fig. 2.1 were (3,3), then (3,3) would be Pareto-inferior to (3,4) in this game because it would be worse for P, and not better for SB, than (3,4). In all the games analyzed in this book, I shall assume that the players can strictly rank outcomes from best to worst, so Pareto-inferior outcomes will be worse for *both* players, as in Fig. 2.1.

itself for a theist if SB is God—but, more problematic for the ratio-
nalist, this outcome is unmistakably worse for both players than
revelation by SB and belief by P, which would confirm P's belief
in SB's existence.[13]

At this point a problem of timing needs to be addressed. Does
SB make his strategy choice first, with P then responding to it by
making his own strategy choice? If so, then there is no paradox,
because SB, realizing the problem that will arise if he chooses his
second (dominant) strategy, would choose his first instead, thereby
inducing P—*after* SB has made his choice—to choose his own first
strategy in order to obtain his best outcome (4), SB his next-best (3).

But I do not think any realistic interpretation of the Revelation Game can so easily resolve the paradox. To begin with, SB
does not obtain his best outcome by choosing first, which his powers discussed in later chapters offer him an opportunity of obtaining. Second, if, in the Revelation Game, P is agnostic, and presumably the kind of player—as opposed to an avowed theist or
atheist—who would take this game seriously, he will play it while
still uncertain about SB's existence. In other words, he will not
know SB's strategy choice, or even whether he exists. For the
agnostic player, therefore, SB's strategy choices might better be
interpreted as SB's establishing his existence or not establishing his
existence, as suggested by his parenthetical strategy choices
shown in Fig. 2.1. (In Section 5.2 I shall consider whether or not
SB's choices are irrevocable when sequential moves are allowed
in the Revelation Game.)

But SB's "not establishing his existence" may occur for two
distinct reasons: (i) he does not in fact exist, or (ii) he does not
choose to reveal himself. The difficulty for P in the Revelation
Game is his inability to distinguish between these reasons. Moreover, as the analysis of the Revelation Game shows, if SB exists, it
is rational for him to choose his second strategy, which deepens

---

13. There is another paradox in the Revelation Game, independent of the
Pareto-inferiority of (2,3), that has been called one of "inducement"; it occurs
because the player without a dominant strategy (P) is induced to make a choice—
by his anticipation that his opponent (SB) will choose his dominant strategy—that
leads to an outcome [(2,3)] ranked higher by the player without a dominant strategy
(P) than the player with one (SB). In other words, the possession of a dominant
strategy hurts one, relatively speaking, in a game like the Revelation Game that is
vulnerable to the inducement paradox. See Nigel Howard, *Paradoxes of Rationality: Theory of Metagames and Political Behavior* (Cambridge, MA: MIT Press), pp.
168–198; and Steven J. Brams, *Paradoxes in Politics: An Introduction to the Nonobvious in Political Science* (New York: Free Press, 1976), Chap. 5, for an analysis of
this paradox and the controversy surrounding it.

the quandary of P in deciding whether SB actually exists or chooses (rationally) not to reveal himself.

In choosing his second strategy of not revealing himself, then SB simultaneously casts doubt on whether he even exists as a player, thereby reinforcing P's agnosticism. On the other hand, SB's choice of his first strategy raises no such problems because SB's "revealing himself" is at least sufficient, and perhaps necessary, for "establishing his existence." These two statements, however, would not be equivalent for the person who, contemplating the "laws of nature" and the order in the universe, finds in them an explanation for the existence of a Grand Designer, who may not overtly reveal Himself.

If SB's strategies need clarification, so, perhaps, does the meaning of "play" of such a game. In a sense, a thoughtful agnostic plays the Revelation Game all his life, never certain about SB's strategy choices—or even that he exists—and wavering between his own. Indeed, we might interpret the agnostic's second strategy—"don't believe in SB's existence"—as an affirmation of his agnosticism. This strategy, as indicated in Section 2.2, is also meant to convey "don't believe in SB's nonexistence"; conjoined, "don't believe in existence/nonexistence" (i.e., be agnostic) is quite different from "believe in SB's nonexistence."

The latter strategy would not be a viable choice for a true agnostic—the P whom I conceive to be the player in the Revelation Game—for whom, presumably, disconfirming evidence, including nonrevelation, would almost never be conclusive. Others, however, do not share this opinion, at least in the case of God. For example, Hanson thinks that the proper position of the agnostic on the question of God's existence should be one of reasonable doubt; for Hanson, moreover, the evidence is tipped decisively against His existence.[14]

In sum, the Revelation Game for the agnostic seems best viewed as one in which a (possible) SB may either establish or not establish his existence. Since the agnostic's choice of believing or not believing in SB's existence was assumed to be independent of SB's strategy choice, the Revelation Game might properly be seen as the 2 × 2 game (two players, each with two strategies) of Fig. 2.1, wherein neither player chooses first. Or, to put it differently,

---

14. Norwood Russell Hanson, The agnostic's dilemma, and what I don't believe, in *What I Do Not Believe, and Other Essays*, ed. Stephen Toulmin and Harry Woolf (Dordrecht, Holland: D. Reidel, 1971), pp. 303–308 and 309–331. I am grateful to Raymond Dacey for this citation.

the players' choices are made in ignorance of each other, though perhaps they continue, in different forms, over P's lifetime—unless he becomes, at some point, a believer or nonbeliever. Then the game for him is no longer a serious one; the evidence, so to speak, is in.

Is there any evidence that human beings and some SB are, in fact, enmeshed as players in the Revelation Game, or that they apprehend their situation in this way? There is none of a factual nature that I know of, though many religions, and the theologies that underlie them, posit some kind of revelatory experience that presumably supports their faith. In the case of an agnostic P, the particular preferences I have postulated for him in the Revelation Game—first a desire for confirmation, with belief preferred to nonbelief—are not, I think, implausible ones to posit. Indeed, postulating as a secondary goal a preference for belief over nonbelief suggests that P is certainly not an inveterate skeptic.

What SB might desire is harder to discern. Certainly the God of the Old Testament very much desired, especially from His chosen people, the Israelites, untrammeled faith and demonstrations of it. Although He never revealed Himself in any physical form, except possibly to Moses before he died, He continually demonstrated His powers in other ways, notably by punishing those He considered transgressors.

If SB, as a secondary goal, preferred revelation over nonrevelation, the (revised) Revelation Game would not be at all problematic. In fact, this preference switch would not even have to apply if P did not believe; as long as P believed (i.e., chose his first strategy), this revision in SB's preferences would transform the outcome (3,4) into (4,4) and (4,2) into (3,2). Then, in this revised Revelation Game, (4,4) would be the evident rational outcome, even though not the product of dominant strategy choices by the players.

The pathology in the Revelation Game in Fig. 2.1 arises because of the reluctance of SB to reveal himself and thereby verify his existence. If some religions did not paint God as all-knowing yet unknowable—and presumably wanting to remain so—there would be no paradox in their tenets of the kind mirrored in the Revelation Game, in which God is SB.

In Section 2.4 I shall discuss why SB may wish to appear inscrutable, and hence why faith must replace reason. Ironically, it is reason, and the logic of game theory, that help to throw into bold relief this tension between faith and reason. Insofar as this tension is captured by the anomaly of a Pareto-inferior rational outcome in the Revelation Game, associated with nonrevelation by SB and nonbelief by P, it alerts us to a problem that belief in SB entails.

2.4. The Knowability Games

A possible sequel to the Revelation Game, assuming SB's existence has been established through revelation of some kind, is what I call the Knowability Games. These games are suggested by a colloquy between Moses and God in the Book of Exodus. When Moses asks God,

> When I come to the Israelites and say to them "The God of your fathers has sent me to you," and they ask me, "What is His name?" what shall I say to them? *(Exod. 3:14)*

God replies,

> "Ehyeh-Asher-Ehyeh [I Am That I Am]." He continued, "Thus shall you say to the Israelites, 'Ehyeh [I Am] sent me to you.'" And God said further to Moses, "Thus shall you speak to the Israelites: The LORD, the God of your fathers, the God of Abraham, the God of Isaac, and the God of Jacob, has sent me to you:
>
> This shall be My name forever,
>
> This My appellation for all eternity." *(Exod. 3:14–15)*

Why would God want to appear so enigmatic? Though His elliptic response to Moses's question makes Him sound as if He will let absolutely nothing out, just three chapters later God divulges His aims with utmost clarity:

> I am the LORD. I will free you from the burdens of the Egyptians and deliver you from their bondage. I will redeem you with an outstretched arm and through extraordinary chastisements. And I will take you to be My people, and I will be your God. And you shall know that I, the LORD, am your God who freed you from the labors of the Egpytians. *(Exod. 6:6–7)*

Thus, the biblical God would appear ambivalent: tight-lipped about His identity, garrulous about His grand design against Pharaoh and the gratitude He expects from the Israelites for abetting their freedom.

In the Knowability Games, I assume that SB has two strategy choices:

1. Be knowable
2. Be unknowable

The two strategy choices of P that I posit are to investigate SB or not, depending on what expectations he harbors about SB:

1. Investigate SB because expect him to be knowable
2. Don't investigate SB because expect him to be unknowable

These expectations can be interpreted as beliefs ("Believe SB to be knowable/unknowable"), but "belief" carries the connotation that something has already happened—giving one reason to believe or not—whereas "expectation" connotes anticipation, a prediction one is willing to make.

In my opinion, a person is more likely to have an *expectation* about SB's character or behavior (e.g., his knowability or unknowability), but to have a *belief* about his existence, though I do not claim this distinction is an airtight one. Rather, it seems to me to be useful in practical discourse.

As for goals, if SB's preferences parallel God's in the Bible, it seems reasonable to postulate the following:

SB: (1) Primary goal—wants to be unknowable, so P's expectation of his unknowability is confirmed, of his knowability is disconfirmed
(2) Secondary goal—prefers that P not investigate him because P expects him to be unknowable

Thus, SB prizes above all else the mystery of his actions and behavior: "You are indeed a God who concealed Himself" (Isaiah 45:15). Not only does he want to thwart a benighted P's unmasking of him, but he would also prefer that P, like Isaiah, expect concealment: "The unexpected is the gift of God."[15]

I assume P has three possible sets of goals, which may be summarized as follows:

I. *He wants to be right:*
(1) Primary goal—wants to confirm SB's knowability/unknowability
(2) Secondary goal—prefers that SB be knowable
II. *He wants SB to be knowable:*
(1) Primary goal—wants to confirm SB's knowability, disconfirm his unknowability

15. This idea is expressed in a novel by Petru Dumitru, *Incognito* (London: William Collins and Sons, 1964), p. 453; its implications and relation to other work are discussed in Rustum Roy, *Experimenting with Truth: The Fusion of Religion with Technology, Needed for Humanity's Survival* (Oxford: Pergamon, 1981), pp. 55ff.

(2) Secondary goal—prefers to investigate SB because he expects knowability

III. *He wants SB to be unknowable:*
  (1) Primary goal—wants to confirm SB's unknowability, disconfirm his knowability
  (2) Secondary goal—prefers not to investigate SB because he expects unknowability

The games that each of these different sets of goal assumptions give rise to are shown in Fig. 2.2. I use the words "confirmed/disconfirmed" to describe the four outcomes, whether P investigates or not, though clearly P's investigation of SB would add certitude to the confirmation/disconfirmation process. In each game, dominant strategies and Nash equilibria (if any) are indicated. Consider the game-theoretic implications of each set of assumptions:

I. SB triumphs; P is in the dark but has not wasted any effort on investigation. This is a game, like the Revelation Game, of partial conflict, with two Pareto-inferior outcomes, (3,1) and (2,2), that neither player desires. Fortunately for the players, unlike the Revelation Game, the Pareto-superior outcome, (4,3), is associated with SB's dominant strategy and is a Nash equilibrium. By confirming SB's unknowability, P is left in the dark, but P's primary goal of being right is fulfilled.

II. SB prevails; P investigates but remains in the dark. In this game, the preferences of the players are diametrically opposed: what is best (4) for one player is worst (1) for the other, and what is next best (3) for one is next worst (2) for the other. Thus, this is a game of *total conflict;* were the players to attach the same values or utilities to their best through worst outcomes (say, 4 through 1, respectively, making the ranks the actual payoffs to the players), it would be called a *constant-sum game.*

This is because the sum of the payoffs to the two players at each outcome is a constant (in this case 5). Were this number subtracted from all the payoffs, the game would be a structurally equivalent *zero-sum game.* In either case, when the game moves from one outcome to another, what one player gains the other player loses, since the sum of payoffs to both players at every outcome remains the same. Thus, wealth or value is neither created nor destroyed.

Both players have dominant strategies in this game, yielding the Nash equilibrium, (3,2). In terms of comparative rankings, it is better for SB (3) than P (2), whose expectation is disconfirmed.

Strictly speaking, though, this is not to say that SB's next-best

Figure 2.2 *Outcome Matrices of Three Knowability Games*

		P	
		Investigate SB because expect knowability (I)	**Don't investigate SB because expect unknowability ($\bar{I}$)**
SB	**Be knowable (B)**	Knowability confirmed	Unknowability disconfirmed
	Be unknowable ($\bar{B}$)	Knowability disconfirmed	Unknowability confirmed

SB: *Wants to be unknowable*

I. *P wants to be right: SB triumphs; P is in the dark but has not wasted any effort on investigation*

		P	
		I	**$\bar{I}$**
SB	**B**	(1,4)	(2,2)
	$\bar{B}$	(3,1)	(4,3) ← Dominant strategy

II. *P wants SB to be knowable: SB prevails; P investigates but remains in the dark*

		P	
		I	**$\bar{I}$**
SB	**B**	(1,4)	(2,3)
	$\bar{B}$	(3,2)	(4,1) ← Dominant strategy

↑
Dominant strategy

III. *P wants SB to be unknowable: Preferences coincidental; both succeed*

		P	
		I	**$\bar{I}$**
SB	**B**	(1,1)	(2,2)
	$\bar{B}$	(3,3)	(4,4) ← Dominant strategy

↑
Dominant strategy

Key: (x,y) = (SB,P)
4 = best; 3 = next best; 2 = next worst; 1 = worst
Circled outcomes are Nash equilibria

outcome is better than P's next-worst, because it is impossible to make subjective comparisons of payoffs involving two different players. Rather, I am simply suggesting that SB "prevails" because he ranks (3,2) higher than P, though in fact the next-best payoff for SB might be a disappointing outcome for him.

III. Preferences coincidental; both succeed. This is a game of *total agreement*, whose Nash equilibrium, (4,4) is the mutually best outcome. It confirms P's expectation of SB's unknowability, which is what SB most desires, too.

There is nothing anomalous about the equilibrium solutions to any of these three games, which illustrate the spectrum of possibilities ranging from total agreement (III) to total conflict (II), with game I falling in between these two extremes. Yet, even the game of total conflict (II) has an equilibrium solution, though such games (and partial-conflict games as well) need not, as will be illustrated later.

It is perhaps noteworthy that SB falls short of his best outcome only when P's primary goal is that he wants SB to be knowable (II), which is in direct conflict with SB's primary goal. While SB can frustrate the achievement of P's primary goal, P can frustrate SB's achievement of his secondary goal, underscoring the compromise nature of this equilibrium solution. Outcome (2,3) in the Revelation Game reflects a similar compromise in goal attainment, but it carries more bite because the players are prevented, by the rules of rational play assumed so far, from attaining the Pareto-superior (3,4) outcome.

To desire knowability and be frustrated in attaining it seems akin to desiring confirmation of SB's existence and being denied the evidence to support one's belief. In each case, P is led to choose a strategy that results in thwarted expectations in confirming knowability in Knowability Games I and II (especially II), or nonbelief induced by SB's nonrevelation in the Revelation Game (Fig. 2.1).

These are not such pleasant outcomes for P or, for that matter, SB. In later chapters I shall show how SB's superiority may attenuate them. But before leaving the Knowability Games, it may come as a surprise to learn that the knowability problem is considerably more baleful if the goals of SB are the following:

SB: (1) Primary goal—wants P not to investigate him because P expects him to be unknowable

(2) Secondary goal—prefers to be unknowable

These goals simply reverse SB's primary and secondary goals

assumed in the previous Knowability Games. They say that SB is less concerned about being enigmatic than having P expect him to be so, even at the risk of having his unknowability disconfirmed.

This supposition about SB's goals receives support from the Bible. While wanting to hide his identity, the God of the Old Testament divulged a great deal about His preferences through both His statements and His actions. His constant admonitions and threats, for example, left little doubt about where He stood, with retribution against miscreants not uncommon. To be sure, God also promised a good deal, as to the Israelites in the passage from Exodus quoted earlier, but His promises were less frequent than His warnings.

Mostly, it seems, God wanted His subjects, especially His chosen people, the Israelites, *not* to try to anticipate His actions before He spoke or acted. At the same time that He wished to appear unknowable, however, He wanted swift and strict obedience to His directives after making His intentions clear.

This is the SB I posit in the fourth Knowability Game in Fig. 2.3. I assume P's preferences are those given earlier in case II: he wants SB to be knowable. The reader may want to check the game-theoretic consequences of case I and III preferences for P,

Figure 2.3 *Outcome Matrix of Fourth Knowability Game (Prisoners' Dilemma)*

SB: *Wants P not to investigate him because P expects him to be unknowable*
P: *Wants SB to be knowable*

		P	
		Investigate SB because expect knowability (I)	**Don't investigate SB because expect unknowability (Ī)**
	Be knowable (B)	Knowability confirmed (1,4)	Unknowability disconfirmed (3,3)
SB	**Be unknowable (B̄)**	Knowability disconfirmed (2,2)	Unknowability confirmed (4,1)	← Dominant strategy

↑ Dominant strategy

Key: (x,y) = (SB,P)
4 = best; 3 = next best; 2 = next worst; 1 = worst
Circled outcome is a Nash equilibrium

when he is of a less inquisitive mind, but I shall consider only the case II preferences because they pose the greatest challenge to *both* players in any of the Knowability Games considered.

As shown in Fig. 2.3, SB and P each has a dominant strategy: SB—be unknowable; P—investigate SB because expect knowability. The intersection of these strategies gives Pareto-inferior outcome (2,2), the next-worst for both players. Thereby, SB's primary goal of wanting P not to investigate him because P expects him to be unknowable, and P's primary goal of having SB be knowable, are frustrated.

Yet, the possibility of achieving these goals by the players' selection of strategies associated with the (3,3) *Pareto-superior* outcome, which is an outcome that cannot be improved by both players simultaneously [(4,1) and (1,4) are better for only one player], seems difficult. Specifically, (3,3) seems to be precluded by the fact that B and $\bar{I}$ are *dominated* strategies—lead to worse outcomes for both players than dominant strategies $\bar{B}$ and I, respectively, whatever strategy the other player chooses. Moreover, (3,3) is not a Nash equilibrium, whereas (2,2) is.

It turns out that the Knowability Game of Fig. 2.3 is game theory's most famous game—Prisoners' Dilemma—and its paradoxical equilibrium outcome has spawned a large literature over the past generation.[16] It is a *symmetrical game:* the players rank two outcomes the same, (2,2) and (3,3), whereas the other two outcomes, (1,4) and (4,1), are mirror images of each other. Later I shall show how SB's omnipotence as well as immortality may be used to counter the choice of (2,2) in this pathological game, but here it is worth mentioning how this outcome differs from the Pareto-inferior (2,3) outcome in the Revelation Game.

The (2,3) outcome in the Revelation Game is associated with the dominant strategy of only one player (SB), whereas (2,2) in the fourth Knowability Game is associated with the dominant strategies of both players. Furthermore, the latter game is symmetrical, so the harm it causes one player—primary goal subverted—is duplicated for the other. By contrast, one player in the Revelation Game (SB) attains his primary, though not his secondary, goal.

---

16. For a description and analysis of Prisoners' Dilemma, see Anatol Rapoport and Albert M. Chammah, *Prisoners' Dilemma: A Study in Conflict and Cooperation* (Ann Arbor, MI: University of Michigan Press, 1965); and Brams, *Paradoxes in Politics*, Chaps. 4 and 8.

Putting aside for now the strategic problem that Prisoners Dilemma raises about rational choice, what does it mean to have "knowability disconfirmed" in the fourth Knowability Game (and also in Knowability Game II in Fig. 2.2)? One interpretation of SB's dominant choice of "be unknowable" is that he would act in such a way as to confuse P. For example, he could try to appear arbitrary by choosing his actions randomly; P could do no better than choose his own dominant strategy of "investigate SB," though his expectation of SB's knowability would be dashed. If P did otherwise, he would risk the choice of (4,1), and similarly SB would risk the choice of (1,4) if he forsook his own dominant strategy.

While these strategies are rational, SB's, in particular, does not seem well designed to teach P any "lessons," especially moral ones that would encourage ethical behavior (however defined). Indeed, random actions that SB intentionally takes to make himself unknowable, without any underlying moral purpose made evident, seem to be the antithesis of what most theists would impute to a benevolent God.

This problem in *all* the Knowability Games is that $\overline{B}$ is a rational choice on the part of SB, even when it leads to his own next-worst outcome (as well as P's in the fourth game, when P makes his own rational choice of I). If SB is equated with God, then the arbitrary or random behavior it is rational for Him to pursue may be impugned for its immorality—it has no apparent rhyme or reason—but it cannot be indicted for being contrary to achieving His ends (as so far assumed).

This sets up a confrontation between rationality and the *problem of evil*—how can a presumably benevolent God condone evil (act arbitrarily) in the world? Of course, if being incognito overrides all other considerations, including ethical ones, there is no inconsistency in SB's dispensing rewards and punishments erratically, without moral purpose. I shall return to the problem of reconciling arbitrary behavior with ethics in Chapter 6, where I discuss qualitatively different kinds of random strategies it may be rational for SB to adopt, depending on his goals, that mix believing and not believing in statements made by P.

Next I want to explore Pascal's decision-theoretic argument for believing in God, which does not posit God as a game player but which P might invoke to justify the belief strategy *not* sustained in the Revelation Game. Then I shall consider related decisions that, like Pascal's wager, presuppose that particular "states of nature" arise in some random fashion.

2.5. *Belief in an Uncertain World*

In his *Pensées*, published posthumously in the late seventeenth century, Blaise Pascal assumed a person is in a betting situation and must stake his destiny upon some view of the world.[17] Starting from the agnostic assumption that "if there is a God ... we are incapable of knowing what He is, or whether He is," and "reason can settle nothing here ... a game [!] is on," Pascal proceeds to invoke reason to say that a prudent person, in his cosmic ignorance, should bet his life on God's existence (Christianity—Roman Catholicism—in Pascal's case). If one does, then the two possible outcomes that may occur have the following consequences:

1. God exists: one enjoys an eternity of bliss (infinite gain)
2. God does not exist: one's belief is unjustified ("loss of nought"), or, at worst, one is chagrined for being fooled (finite loss)

Since outcome 1 promises an infinite gain, whereas outcome 2 leads to only a finite loss, the choice seems clear to Pascal: one should believe in God's existence.

This argument, made in No. 223 of *Pensées*, is, however, quite muddled. It implies that belief in God is justified by the infinite expected payoff (defined below) alone. However, in choosing between belief and nonbelief, it is proper to compare the expected payoffs that each of these alternative courses of action yields and choose the higher. Pascal is not explicit about this, but the calculation of expected payoff for nonbelief described below is suggested by Pascal's discussion of the "fear of hell" in No. 227 of *Pensées*.

Pascal's argument, as restated above, holds even if it is not true, as Pascal assumed, that "the chances of gain and loss are equal." For, however small (though positive) the chances of outcome 1 are, when multiplied by an infinite gain, the resulting *expected payoff* (sum of utilities of outcomes times their probabilities of occurring) for believing in God's existence is infinite. Since an infinite payoff has no obvious meaning, one might think of this payoff as some stupendous, but finite, reward.

By comparison, if one bet that outcome 2 (God does not exist) were true, but it turned out to be false, one would suffer an eternity of torment (infinite loss)—or a colossal penalty. (Henceforth,

---

17. In quoting from *Pensées*, I use the arrangement and numbering in *Pascal's Pensées*, translated by H. F. Stewart (New York: Pantheon, 1950); all citations in the text are from No. 223.

stupendous rewards and colossal penalties might be substituted for infinite gains and infinite losses to make this discussion more concrete—and mathematically acceptable since, technically, one would evaluate a player's expected utility in the limit as the reward/penalty approached infinity.)

In short, believing in God's existence yields an infinite positive expected payoff (infinite gain minus finite loss), whereas not believing yields an infinite negative expected payoff (infinite loss minus finite gain). Since this calculation can be made before the outcome is known (if it ever is!), it suggests that P's key to happiness—and perhaps heaven—is not revelation, as in the Revelation Game when P believes and SB reveals himself, but being prepared for the *possibility* that God exists by believing in Him.

Even if one should not attain eternal bliss, at least it can be said that one made an honest effort in this direction. Furthermore, Pascal avers that the very act of believing in this calculated fashion sets up the conditions for developing genuine faith and becoming a true believer, so there is nothing insincere or immoral about starting off by making an expected payoff calculation, though one's faith is ultimately sustained "by taking holy water, by hearing mass, etc."—the accoutrements of religion.[18]

One weakness, I believe, in Pascal's wager argument is that he never postulated a third possibility (or still others): SB's existence is indeterminate—information that would settle this question is unattainable. I would like to add this as a possible *state of nature*, or a situation that might arise by happenstance in the world, in a somewhat different Search Decision (to distinguish it from a game in which the players make deliberate choices).

The three states of nature that I assume P can perceive are: (i) SB's existence is verifiable, (ii) SB's nonexistence is verifiable, (iii) SB's existence/nonexistence is indeterminate. True, the third state may in fact hide one of the other two, but P may simply not be capable of verifying SB's existence/nonexistence. Hence, *in P's perception*, the third state and the other two are mutually exclusive: observing or experiencing (iii) precludes observing or experiencing (i) and (ii), just as the latter two states preclude each other and (iii).

It is perhaps strange that Pascal did not postulate this third

18. For an illuminating discussion of some of the problems of "going through the motions" of believing in order to generate the real thing, see Jon Elster, *Ulysses and the Sirens: Studies in Rationality and Irrationality* (Cambridge: Cambridge University Press, 1979), pp. 47–54.

state in his wager, because he said that God, being "infinitely incomprehensible," forecloses the possibility that His existence can ever be determined. Pascal's wager, in other words, presumes a bet whose outcome will never be known—at least in one's present life.

To put it somewhat differently, Pascal seems to have thought we will know in our lifetimes only indeterminacy. Nevertheless, he postulates the other two more determinate states, presumably because he thinks we might learn of the existence of God in the hereafter. I am not so sure this is the case and, therefore, prefer to retain the third state in present calculations.

The three postulated states in the Search Decision are shown in Fig. 2.4, along with P's strategies of searching (S) and not $(\overline{S})$ for SB. I interpret S to mean that P tries to learn something more about SB, including his existence and what form it takes, and $\overline{S}$ to mean he makes no effort to understand the Great Mystery, if you will.

The six payoffs associated with this decision are ranked from best to worst for P. Presumably, P would most like the search to verify SB's existence (6) and least like it to fail (1). In between, I assume, P would prefer to search even if he only succeeds in being able to verify SB's nonexistence (5) or there is no information and his failure to search is appropriate (4); he would less prefer not to search when there is information that would make SB's nonexistence (3) or existence (2) verifiable. Although Pascal would probably rank this last outcome worst, switching 2 and 1 in Fig. 2.4, this

Figure 2.4 *Outcome Matrix of Search Decision/Game*

		States of Nature		
		i. SB's existence verifiable	*ii. SB's non-existence verifiable*	*iii. Indeterminate (no information)*
Search (S)		Search successful (6)	Search productive (5)	Search unavailing (1)
P				
Don't search (S̄)		No search unwise (2)	No search wise (3)	No search appropriate (4)

No dominant strategy for P

No dominant strategy for SB (in a game)

Key: 6 = best; ... 1 = worst

alternative ranking would not affect the argument in the next paragraph.

Whatever utilities one attaches to the outcomes consistent with the ranks in Fig. 2.4, P does not have a dominant, or unconditionally best, strategy: S is better than $\overline{S}$ if SB's existence or nonexistence is verifiable, but $\overline{S}$ is better than S if the situation is indeterminate—there is no information to be had. This last state, incidentally, complicates P's choice, because P's ordering of $\overline{S}$ (4) over S (1) in this state is different from his ordering of the other two states, in which he prefers S over $\overline{S}$.

Thus, P would be in a quandary, based on his presumed ordering of the possible outcomes, because his two strategies are *undominated:* neither is unconditionally best, or dominant, nor unconditionally worst, or dominated. Moreover, one cannot sidestep the problem as easily as Pascal did, because P's best *and* worst outcomes are both associated with S. If infinite gain (eternal bliss) accrues to P from the former, and infinite loss (eternal torment) from the latter, then the expected payoff (which would sum positive infinity and negative infinity) when P chooses S is not defined, rendering it truly imponderable.

Whatever the (nonzero) probabilities associated with each state of nature are, therefore, one cannot say whether the expected payoff of searching is greater than that of not searching. So again P would be in a quandary if his ordering of outcomes were that in Fig. 2.4, though a switch of 2 and 1, and the assignment of infinite positive and negative payoffs to P's best and worst outcomes, respectively, would throw the decision in favor of S.

For the sake of argument, assume that the states of nature do not occur by chance, according to some random mechanism, but that SB can choose which will arise, or at least induce P to think one has occurred. That is, SB can choose a strategy that would *indicate* that his existence or nonexistence is verifiable, or that his existential state cannot be determined, which is not the same as deciding whether to exist or not. For example, SB could indicate state (i) by signs of revelation, or state (iii) by evidence of nonrevelation. [How SB would indicate that his nonexistence can be verified in state (ii) is not so evident.]

What state will SB choose? Like P, he has no dominant strategy in this decision-transformed-into-a-game if his ordering of the six outcomes duplicates P's. But a mutually best (6,6) outcome would almost surely induce both players to choose their strategies associated with it in this game of total agreement. Yet, as I argued in the case of the Revelation Game, SB would seem to have good reason *not* to make his existence apparent, though he might want to signal that state (i), and maybe state (ii), should not be ruled out in order to push P in the direction of searching. This may induce P to be obedient, too, though to what is not clear.

For most agnostics, I presume, the signals, if they hear or see any, are ambiguous. This probably predisposes them to think indeterminate state (iii) is most probable, and hence their search should not continue, unless at the same time they associate very high utility with being successful or productive in their search [in (improbable) states (i) and (ii)].

I shall not speculate further on P's choices in this situation, or on SB's choices if he in fact has them in a game, wherein he can choose the states of nature. But it should be noted that P's apparent certitude in favor of belief in Pascal's wager founders in the Search Decision when one asks much less of P: Should he continue the search for a key to unlock the Great Mystery, or is his effort likely to go unrewarded? As I have shown, answers depend on what P views to be the states of nature, how he orders his choices with respect to them, and sometimes what probabilities he attributes to the states and what utilities he associates with the outcomes.

This will be an unsatisfactory response to those who seek definite answers, not further qualifications. Yet, these qualifications cannot be divorced from rational choices, though one might contend that belief in SB or God is not, or cannot be, a rational choice. William James, for example, maintained that our beliefs ought to be determined by "our passional nature" when their truth cannot "be decided on intellectual grounds."[19]

But "passions," like preferences and goals, are not rational or irrational—choices based on them are. Though passions tend to be more identified with emotions than logic, to be "driven" by one's passions is in fact tantamount to acting to try to satisfy one's preferences and achieve one's goals. Even if one cannot articulate a logic to one's passions, is it not rational to try to satisfy them?

To make this point more concrete, consider the Concern Decision shown in Fig. 2.5. Whether it is interpreted as a "logical decision" or an "emotional commitment" to be concerned about SB, the rational calculus is the same. To be sure, passions may dictate different assignments of utilities to the outcomes, as I shall illustrate below, but this has no bearing on making rational choices that mirror these passions as closely as possible.

19. William James, The will to believe, in *The Writings of William James*, ed. John J. McDermott (Chicago: University of Chicago Press, 1967), p. 723.

Implicit in the choice that P faces is the supposition that SB exists, but P does not know whether SB is aware, or cares, whether P is concerned about him or not. If SB does care, I assume P prefers to be concerned—that is, tries to take into account SB's preferences in choosing his own (P's) actions. (Here an emotional interpretation might be that P is empathetic, a strategic interpretation that P knows SB will know his choice and hold him responsible, so it "pays" to be pious.)

If SB does not care, I assume P is better off not concerning himself with SB's preferences, for this is simply wasted effort (even for the empathetic P, for SB in this case is not worthy of empathy). Note that I make no assumptions about what SB's preferences in fact are, only that P may or may not show concern for them.

Assume, further, that P most prefers to be concerned when SB cares (4 in Fig. 2.5) (justified concern); next best (3) is when P is unconcerned and SB does not care (justified unconcern). These outcomes, by the assumptions in the previous two paragraphs, rank higher than concern/uncaring (2) and unconcern/caring (1). I assume that the latter outcome is P's worst because he abandons an SB who cares, which in Pascal's scheme of things would lead to unmitigated agony.

Even though P's choice here—to be concerned or not about SB—is not exactly the way Pascal presented the alternatives in his wager, Pascal, I imagine, would advise as follows: be concerned, for there is an infinite reward associated with rank 4, an infinite penalty with rank 1, whereas the payoffs associated with ranks 2 and 3 are only finite so do not matter in the expected-payoff calculation. For whatever the probabilities associated with each state

Figure 2.5 *Outcome Matrix of Concern Decision*

| | States of Nature | |
	SB aware/ cares whether P is concerned	*SB not aware/ does not care whether P is concerned*
Be concerned	Concern justified (4)	Concern unjustified (2)
P		
Be unconcerned	Unconcern unjustified (1)	Unconcern justified (3)

No dominant strategy for P

Key: 4 = best; 3 = next best; 2 = next worst; 1 = worst

of nature, as long as they are positive there is an infinite gain associated with being concerned, an infinite loss with not being concerned.

If one does not use these utilities, however, but considers only P's ranking of the four possible outcomes, P would be in a dilemma. Without a dominant strategy, his best choice depends on what state of nature arises.

Most theists, I presume, would assert that God is concerned, whereas atheists would say that the question is meaningless because God does not exist. Thus, because of their different "passional natures"—believers with a passion for a concerned God, nonbelievers with a passion against any God—neither would have a problem about which choice to make in the Concern Decision. For agnostics, on the other hand, being concerned or unconcerned are undominated strategies, based on P's assumed ranking of the four outcomes.

Pascal, I believe, would probably accept this ranking. He could, as I suggested earlier, resolve the dilemma by assigning infinite positive utility to rank 4, infinite negative utility to rank 1, but, practically speaking, what do these huge rewards and penalties signify? They might be considered a gauge of P's passions—very strong—but are they meaningful if P thinks that the probability of the first state (SB aware/cares) is negligible? If SB is much more likely to be unaware or indifferent, perhaps these attributions of value should be ignored in favor of the choice between the middle rankings in the second state (SB not aware/does not care), throwing P's decision in favor of being unconcerned.

One possible way around the dilemma would be to alter P's ranks, switching, say, 4 and 2 or 4 and 1. In the former case, being concerned would be a dominant strategy; in the latter case, being unconcerned would be dominant. However, I find it difficult to justify these new assignments, which say that P's best outcome is to be concerned when SB does not care, and to be unconcerned when he does, reversing the attitudes it seems to me reasonable to impute to P. These switches smack of playing with numbers to sidestep a genuine dilemma; in my opinion, they are quite absurd.

What may be more defensible is to switch 3 and 4, arguing that justified unconcern, which might save P much time and effort, is better than justified concern. After all, either expression—of concern or unconcern—is borne out by the "facts" (i.e., the state of nature). Conceding that the claim that the present 4 outcome in Fig. 2.5 offers more to P than the present 3 outcome is contestable, what would be the effect of interchanging 4 and 3 in the Fig. 2.5

outcome matrix? Or of switching the 2 and 1 ranks that indicate unjustified concern/unconcern, based on a similar argument that to be concerned is more demanding than being unconcerned (even unjustifiably)?

Unfortunately for P, neither the 4-3 nor the 2-1 switches, or even both together, would endow him with a dominant strategy. The dilemma would remain: his best choice, at least based on the rankings, would depend on the state of nature that occurs.

This dilemma also characterizes the Search Decision. To see this, simplify this decision by removing either state of nature (ii) or (iii) shown in Fig. 2.4. Because P's best outcomes lie along the main diagonal of the matrix with either state deleted, and his worst outcomes along the off-diagonal, he does not possess a dominant strategy. Moreover, no switching of either diagonal or off-diagonal ranks alters this fact.

To return to the Concern Decision, one's passions—such as those I presumed of Pascal for heaven and against hell—may offer relief by throwing the choice one way or another. But the rankings by themselves that I presumed do not.

For the agnostic who has these preferences, the absence of an obviously best choice, not dependent on what state of nature arises, poses a problem. This, I think, is what the decisional calculus I developed in this section helps to clarify, even if it does not provide any resolution to the resulting indecision, and maybe anxiety, P faces in the situations depicted here.

2.6. *Conclusions*

In this chapter I have analyzed problems that inhere in two different kinds of theological choice situations: games, in which SB is an active player, like P, with preferences, who desires to attain his best possible outcome; and decisions, in which states of nature arise, according to some chance mechanism, that characterize SB's presence or interests, but in which P alone makes choices without ever being certain of what state will in fact arise.

In both games and decisions—which might be thought of as one-person *games against nature*—the players are assumed to make rational choices with respect to their preferences or goals. In general, this means choosing a dominant strategy if one has one; anticipating the other player's choice of such a strategy if one does not have one; or making a choice that maximizes one's expected payoff. The latter choice takes into account, at least in a rough way, the players' utilities for outcomes (only finite and infinite so far, but more precise calculations will be made later) and the probabilities that particular states of nature occur. The Revelation Game (Fig. 2.1) shows, for plausible assumptions about P's and SB's rankings of outcomes, that SB has a dominant strategy of nonrevelation, which induces P not to believe in his existence. The resulting outcome is Pareto-inferior—worse for both players than revelation by SB and believing by P—as is the outcome in the fourth Knowability Game (Fig. 2.3).

In the latter game, which is a Prisoners' Dilemma, it is rational for SB to be unknowable and P not to investigate SB because he expects unknowability. Insofar as SB's mysterious behavior to achieve this end entails acting arbitrarily, it would appear devoid of moral purpose and hence loathesome. This game, then, suggests that the problem of evil might be an anomalous product of the strictures of rationality, which dictate purposeless rewards and punishment to prevent one from becoming knowable.

Other difficulties crop up in the Search Decision and Concern Decision, in which P's choice of searching for, or being concerned about, SB depends on the verifiability of SB's existence/nonexistence, or on SB's awareness and caring for P. Pascal's neat solution to this intellectual imbroglio was to suppose payoffs to P that presumed a halcyon heaven and a horrible hell. Thereby he justified belief in God, which the Revelation Game did not support, as infinitely rewarding, on the average, even if this belief turned out to be false.

Pascal's payoff assumptions may strike agnostics as loading the dice to prop up a result that Pascal felt compelled to find intellectual arguments to support. Yet, as I tried to show, "our passional nature," in James's felicitous phrase, is only part of the story: attributions of value, or utility, are neither rational nor irrational. Choices that lead to outcomes yielding payoffs may or may not be; the problem is that the games and decisions discussed in this chapter generate genuinely paradoxical rational choices, or pose vexing dilemmas, for plausible assignments of passions/preferences to the players.

Is the problem of making rational and coherent choices in theological matters of this sort insoluble? Before passing judgment on this question, it behooves us to consider what effects SB's superior qualities might engender in these and other trying situations.

THREE

Omniscience and Partial Omniscience

3.1. Introduction

The picture presented in Chapter 2 is a bleak one, at least for the questing agnostic searching for more than straws in the wind to ward off doubt and uncertainty and justify his beliefs/expectations. If the Revelation Game accurately represents his preferences as well as SB's (or a possible God's), the strategy choices are clear: SB would not reveal himself, or establish his existence, and P, anticipating SB's dominant strategy choice, would not believe in his existence. Hence, P would presumably remain an agnostic, which reverses the rationalistic faith argument Küng sets forth in *Does God Exist?*[1]: it is in fact rational, if one is playing the Revelation Game, not to believe in SB's existence. Recall, though, that this strategy choice does not imply that it is rational for the agnostic to believe in SB's nonexistence, a strategy choice not available to P in the Revelation Game as I interpreted it, because SB could exist without revealing himself.

---

1. Hans Küng, *Does God Exist? An Answer for Today,* translated by Edward Quinn (New York: Doubleday, 1980).

The fact that the apparent rational strategy choices of the players in the Revelation Game lead to an outcome inferior for both to another outcome in the game is particularly distressing. This pathology, as I showed in Section 2.3, can be remedied by assuming that SB chooses his strategy first, and P responds to his choice, but this assumption seems hard to support by a plausible scenario that describes how this game might be played by an intelligent agnostic. Moreover, SB, as I shall show later, can do still better in the Revelation Game by invoking his superior powers.

Letting SB make the first move in the fourth Knowability Game (Fig. 2.3), by contrast, offers the players no relief from the Pareto-inferior rational outcome [(2,2)] in this game. For if SB chooses to be knowable, at the start, P's best response is to investigate SB; because P can realize his best outcome, and SB's worst [(1,4)], by expecting knowability—an expectation, of course, confirmed by SB's investigation—P will expect it.

This expectation by P would obviously sabotage the possibility of the Pareto-superior (3,3) "cooperative" outcome in this game. Similarly, P would have no incentive to expect unknowability if he chose first; so allowing either player the first move will not enable them to extricate themselves from this Prisoners' Dilemma.

Although the clash of goals of the players in this game seems to foredoom their fate, in this chapter I shall show how "partial omniscience," coupled with some other conditions, may offer relief. To carry out this analysis, I shall demonstrate that a recent problem in philosophy that posits an almost omniscient superior being, known as "Newcomb's problem," can be reformulated as a Prisoners' Dilemma in which both players have considerable, if not perfect, predictive powers.

Next, using expected-payoff calculations of the kind discussed in Section 2.5, I shall show how decision-theoretic analysis can be incorporated in games, like Prisoners' Dilemma, that seem to pose hopeless difficulties for the players. Different models, which incorporate other aspects of omniscience as well as omnipotence and immortality, will be applied to Prisoners' Dilemma and to similar intractable games in later chapters to determine what, if any, comfort they may bring to one or both players.

To begin the analysis in this chapter, I will show that a straightforward definition of omniscience has no effect on the outcomes of any of the games discussed in Chapter 2 if SB possesses this predictive power. Curiously, if P possesses this power, it helps the players in the Revelation Game, though not in any of the Knowability Games. After discussing Newcomb's problem and its connection to Prisoners' Dilemma, I shall return to these games at the end of the chapter.

3.2. *Some Effects of Omniscience*

In Section 2.2, I argued that SB is not likely to make the first move in the Revelation Game, even though he could induce outcome (3,4), his next-best outcome. Among other reasons, if he resembles the biblical God, making his presence manifest would obviate any test of faith of P. Likewise, P would probably not anticipate such a move because, as an agnostic, he would presumably have doubts about SB's existence or, if he thought SB were possibly a player, would understand as well as SB that SB would invalidate any test of faith if he gave away his presence through revelation. At the same time, P might reason that if SB were a player, he would await his (P's) move, or, if SB possessed omniscience, could foretell his move anyway and then act on this foreknowledge.

To develop the implications of this idea of omniscience, assume that the latter is the case: SB is *omniscient* and, as a consequence, can predict P's strategy choice before the game is played.[2] In addition, assume that P knows that if SB exists, he is omniscient and will act on the basis of his prediction of P's strategy choice to ensure his best possible outcome. Do these assumptions about SB's omniscience and P's knowledge of it change what rational players would choose in the Revelation Game (Fig. 2.1)?

The answer, surprisingly, is no. Because SB has a dominant strategy, the fact that he can predict P's strategy choice will not enable him to do any better: his best choice, whichever strategy P chooses, remains his dominant strategy (not to reveal himself) because, by definition, it is unconditionally best. Hence, having advance information on P's strategy choice does not help SB an iota in the Revelation Game—the pathological (2,3) outcome remains the rational choice of the players. This constancy, however, does not obtain in all games, wherein a player may either be helped or hurt by omniscience and by his opponent's awareness that he possesses this capability, as I shall show in Chapter 4.

Now, if P were the omniscient player able to predict SB's

---

2. An interesting side issue: Does SB's omniscience abrogate P's free will? Probably not, because SB's accurate prediction only implies that the predictor has discovered causes that underlie P's choices; it does not follow that he forces them and thereby prevents P from choosing freely.

choice of a strategy before play of the Revelation Game, and if SB knew this, then it would be rational for SB to establish his existence by revealing himself. The reason is that P, predicting SB's revelation choice, could ensure not only his best outcome but also a better outcome for SB, (3,4), by believing in SB's existence and having it confirmed when SB actually chooses (according to P's prediction) to reveal himself.

Endowing P with omniscience is equivalent to assuming that SB moves first, for in each case P would know SB's strategy choice before he made it. Giving P omniscience, however, seems no more credible than giving SB the first move, which is an assumption I rejected earlier; so I think the salutary consequence these assumptions have for the players, by inducing the (3,4) outcome in the Revelation Game, is difficult to sustain.

SB also does not benefit from being omniscient in any of the four Knowability Games discussed in Section 2.4. Recall that in each of these games SB has a dominant strategy—be unknowable ($\overline{B}$)—so the fact that he can anticipate P's choice would not in any instance move him from choosing this strategy: $\overline{B}$ is better whatever P does or whatever he predicts P will do. Since P has a dominant strategy himself in three of the four Knowability Games, his possession of omniscience, implausible as this may be, would not help him either. In the remaining game (I in Fig. 2.2), P would anticipate SB's dominant strategy choice of $\overline{B}$, which would only reinforce his own best response of $\overline{I}$, leading to (4,3), the same outcome that would be chosen by nonomniscient players.

In the analysis so far, the nonomniscient player's knowledge of the other player's omniscience is inconsequential in any game except the Revelation Game, and here only when it is SB who has knowledge of P's omniscience. This role reversal induces SB to choose his dominated strategy of revelation; P, having predicted this, can rest assured that his best choice is to believe in SB's existence.

This ascription of omniscience to P makes little sense in the Revelation Game. The more sensible ascription of omniscience to SB, on the other hand, may have paradoxical consequences when P is aware that SB possesses this predictive power, as I shall demonstrate in Chapter 4, though it has no effect in the Revelation Game.

There is no awareness on the part of P in the Search and Concern Decisions discussed in Section 2.5, because SB is not assumed to be a player. Rather, his existence or aspects of him appear as states of nature, and these certainly do not make predictions that

P can anticipate. Yet, insofar as P can make predictions himself about which state of nature will occur, he can benefit in each of these decisions.

Thus, in the Search Decision (Fig. 2.4), P should search if he thinks SB's existence or nonexistence is verifiable, not search otherwise. In the Concern Decision (Fig. 2.5), he should evince concern if he thinks SB cares, be unconcerned otherwise. These are obvious consequences of these decisions, and I mention them only to indicate what omniscience in a game is not.

This concept entails, first and foremost, the ability to predict the choices of another active player—one who may or may not share one's interests. I apply it primarily to a SB and consider the consequences for a P who is aware of the SB's ability. It is not a concept I apply to the prediction of states of nature governed, perhaps, by underlying forces that are not well understood and cannot be predicted with precision. For example, earthquakes seem at any moment to be random events, though better physical theories would surely facilitate predictions of their occurrence.

True, if there is a God, pantheists from Spinoza to Einstein have seen in Him an indifferent natural force that may give rise to phenomena like earthquakes. But their God strove for harmony and simplicity; if the Creator had not imbued the universe with order and coherence, Einstein said, "Then I would have been sorry for the dear Lord."[3]

Emphatically, for Einstein, God does not take a personal interest in our affairs. To understand and make evident His order and coherence justifies search (and reflection), but not human concern for Him as such.

Einstein, in precluding God's role as a game player, nevertheless gave Him a peculiar turn of mind: He is "subtle" but never "malicious."[4] The subtlety in Newcomb's problem that has bedeviled philosophers for the last ten years is the subject to which I turn in Section 3.3.

By way of preview, this problem was formulated by a physicist, William A. Newcomb, in 1960, elucidated by a philosopher,

---

3. Quoted in *Some Strangeness in the Proportion: A Centennial Symposium to Celebrate the Achievements of Albert Einstein*, ed. Harry Woolf (Reading MA: Addison-Wesley, 1980), p. 379. On his religious beliefs, Einstein is quoted as saying (p. 16): "I believe in Spinoza's God who reveals himself in the orderly harmony of what exists, not in a God who concerns himself with the fate and actions of human beings." Spinoza speaks for himself in Benedictus de Spinoza, *Chief Works*, translated by R. H. M. Elwes (New York: Dover, 1951).

4. *Some Strangeness in the Proportion*, p. 480.

Robert Nozick, in 1969,[5] popularized by Martin Gardner in 1973,[6] and—despite the fact that it generated a huge response from many different people—remains, in the opinion of Nozick, a more open problem than ever.[7] After describing this problem in Section 3.3, I shall indicate a persuasive resolution of it, due to John A. Ferejohn, in Section 3.4 that turns on reformulating it as a decision, or a one-person game against nature.

Next, in Section 3.5, I shall show that there is an intimate relationship between Newcomb's problem and Prisoners' Dilemma, the latter game being a "symmetricized" version of the former in its payoff structure. Tying this analysis to games played between SB and P, I shall then demonstrate how the assumption made about SB in Newcomb's problem, when applied to P as well in Prisoners' Dilemma—one player (SB?) considered as a leader and the other (P?) as a follower—offers one resolution of this dilemma. Another resolution, based on the "theory of moves," will be discussed in Chapter 4.

3.3. Newcomb's Problem[8]

Imagine the following situation. There are two boxes, B1 and B2. B1 contains $1,000; B2 contains $1,000,000 or nothing, but you do not know which. You have a choice between two possible actions:

1. Take what is in both boxes
2. Take only what is in B2

5. Robert Nozick, Newcomb's problem and two principles of choice, in *Essays in Honor of Carl G. Hempel*, ed. Nicholas Rescher (Dordrecht, Netherlands: D. Reidel, 1969), pp. 114–146.

6. Martin Gardner, Mathematical games, *Scientific American* (July 1973), 104–108.

7. See Nozick's reply to those who responded to Gardner's *Scientific American* article (note 6) in Martin Gardner, Mathematical games, *Scientific American* (March 1974), 102–108. Since Nozick's reply, the literature on Newcomb's problem has expanded rapidly and is too large to cite here. One recent article of note, with a number of citations, that bears directly on the argument to be made in this chapter, is David Lewis, Prisoners' Dilemma is a Newcomb problem, *Philos. Public Affairs* 8, 3 (Spring 1979), 235–240. See also Isaac Levi, A Note on Newcombmania, *J. Philos.* 79, 6 (June 1982), 337–342; and Ellery Eells, *Rational Decision and Causality* (Cambridge: Cambridge University Press, 1982).

8. Sections 3.3–3.7 are based on Steven J. Brams, Newcomb's problem and Prisoners' Dilemma, *J. Conflict Resolution* 19, 4 (December 1975), 596–612. This material was also used in Brams, *Paradoxes in Politics: An Introduction to the Nonobvious in Politics* (New York: Free Press, 1976), Chap. 8.

Now what is in B2 depends entirely upon what action some SB predicted you would take. If he predicted you would (1) take what is in both boxes—or would randomize your choice between the two actions—he put nothing in B2; if he predicted you would (2) take only what is in B2, he put $1,000,000 in B2. Hence, you are rewarded for taking only what is in B2—provided SB predicted this choice—though you have some chance of getting even more ($1,001,000) if you take what is in both boxes and SB incorrectly predicted that you would take only what is in B2. On the other hand, you do much less well ($1,000) if you take what is in both boxes—and SB predicted this action—and worst ($0) if you take what is in B2 and SB incorrectly predicted that you would take what is in both boxes.

These payoffs are summarized in the payoff matrix of Fig. 3.1. Clearly, the very best ($1,001,000) and very worst ($0) outcomes occur when SB's predictions are incorrect, the intermediate outcomes ($1,000,000 and $1,000) when SB's predictions are correct.

Note that SB's strategies given in Fig. 3.1 are predictions, not what he puts in B2. One could as well define his two strategies to be "Put $1,000,000 in B2" and "Put nothing in B2," but since these actions are in one-to-one correspondence with his predictions about what you take, it does not matter whether we consider SB's strategies to be predictions or actions. (Since SB's predictions precede his actions, they are perhaps the more basic indicator of his behavior.)

From the perspective of game theory, what matters is that SB's strategies are not the "free" choices usually assumed of players in the normal-form representation of a game (i.e., its representation by a payoff matrix). Also, there is no indication of payoffs to SB. But this is not a game in the usual sense, which renders its Fig.

Figure 3.1 *Payoff Matrix for Newcomb's Problem*

		SB	
		Predicts you take only what is in B2	*Predicts you take what is in both boxes*
You	*Take only what is in B2*	$1,000,000	$0
	Take what is in both boxes	$1,001,000	$1,000 ←Dominant strategy

3.1 representation vulnerable to attack (as will be shown in Section 3.4). Moreover, the solution that will be proposed in Section 3.6 to a symmetrical version of this game rests on a different model of player choices.

At first sight, it would appear, Newcomb's problem does not present you with a problem of choice. Your second strategy—take what is in both boxes—dominates your first strategy—take only what is in B2—since whatever SB predicts, your payoffs are greater than those associated with your first strategy. Thus, you should always take what is in both boxes, which assures you of at least $1,000, as contrasted with a minimum of $0 for your first strategy.

This choice is complicated, however, by your knowledge of the past performance of SB, who is (or seems) "superior" precisely because his predictions have always been correct. (SB may be thought of as God here as elsewhere, but the paradox described subsequently retains its full force if he is regarded as a superior intelligence from another planet or a supercomputer capable of discerning your thoughts and using this knowledge to make highly accurate predictions.) Although you do not know what his prediction is in the present choice situation, it will, you believe, "almost surely" be correct. Thus, if you choose your dominant strategy of taking what is in both boxes, SB will almost certainly have anticipated this action and left B2 empty. Hence, you will get only $1,000 from B1.

On the other hand, if you choose your first strategy and take only what is in B2, SB, expecting this, will almost surely have put $1,000,000 in B2, which would seem a strong argument for choosing this strategy, despite the dominance of your second strategy. This argument is based on the principle of maximizing *expected utility*—equated with money in this case—which is the sum of the payoffs associated with each of the outcomes in each strategy times the probability that each will occur. (Recall that a similar calculation was made in Section 2.5, but the payoffs were identified as only "finite" or "infinite," whereas the utilities here are concrete finite quantities.)

Assume, for purposes of illustration, that though you have near-perfect confidence in the predictions of SB, you conservatively estimate that the probability of his being correct is only 0.9. Then, the expected utility of your first strategy (take only what is in B2) is

$$(\$1,000,000)(0.9) + (\$0)(0.1) = \$900,000,$$

whereas the expected utility of your second strategy (take what is in both boxes) is

($1,001,000)(0.1) + ($1,000)(0.9) = $101,000.

Evidently, to maximize your expected utility, you should take only what is in B2. (In fact, the probability that SB is correct need only be greater than 0.5005 in this example to make the expected utility that you derive from your first strategy exceed that which you derive from your second strategy.)

This conflict between the dominance principle, which prescribes taking what is in both boxes, and the expected-utility principle, which prescribes taking only what is in B2, is the heart of the paradox. Although each principle can be supported by very reasonable arguments, the choices that each prescribes will generally be in conflict, given the excellent powers of prediction assumed on the part of SB.

This paradox, it should be pointed out, is not the product of any hidden or suppressed assumptions. You are assumed fully to understand the choice situation, SB knows that you understand, and so on.

Furthermore, no kind of "backwards causality" is assumed to be at work, whereby your present actions influence SB's past predictions. SB is assumed to have made a prediction—say, a week before you make your choice—and put either $1,000,000 in B2 or nothing. The money is there or it is not there, and nothing that you think or do can subsequently change this fact.

What is your choice? Nozick reports:

> ... I have put this problem to a large number of people, both friends and students in class. To almost everyone it is perfectly clear and obvious what should be done. The difficulty is that these people seem to divide almost evenly on the problem, with large numbers thinking that the opposing half is just being silly.[9]

Although respondents to Gardner's first *Scientific American* article favored the expected-utility principle by better than two to one, Nozick concludes in his reply to the respondents that "the [148] letters do not, in my opinion, lay the problem to rest."[10]

9. Nozick, Newcomb's problem and two principles of choice, p. 117.

10. Nozick's reply in Gardner, Mathematical games, *Scientific American* (March 1974), 108.

3.4.   Which Principle, and Is There a Conflict?

Is there any solution to this paradox that resolves the apparent inconsistency between the dominance principle and the expected-utility principle? John A. Ferejohn has shown that if Newcomb's problem is reformulated as a *decision-theoretic* rather than a *game-theoretic* problem, the apparent inconsistency between the two principles disappears.[11]

From Section 2.5 recall that in a model of decision making under uncertainty, the action an actor takes does not lead to a particular outcome with certainty but to a set of possible outcomes with, sometimes, specific probabilities of occurrence. Conceptualized in these terms, the person making the choice of either B2 or both boxes in Newcomb's problem does not view SB as making predictions *about what he will choose* but rather making predictions *that are correct or incorrect* (see Fig. 3.2).

Observe that your two best outcomes in the payoff matrix of Fig. 3.1 ($1,000,000 and $1,001,000) are both associated with SB's prediction that you will take only what is in B2 (first column of Fig. 3.1). By contrast, in the decision-theoretic payoff matrix of Fig. 3.2, these outcomes are the diagonal elements, each being associated with a different state of nature, which is assumed to be either a correct or an incorrect prediction on the part of SB. Because your best choice depends on what state of nature occurs in the decision-theoretic representation (if SB is correct, take only what is in B2; if

11. Personal communication, May 27, 1975. Nigel Howard has also shown these two principles to be consistent in a metagame representation of Newcomb's problem (see note 13 for references to metagame theory). Personal communications, March 27, 1975 and June 25, 1975. Whereas Howard's metagame resolution of the paradox retains the assumption that Newcomb's problem is a game, Ferejohn criticizes precisely this assumption, as I shall show forthwith.

Figure 3.2   *Newcomb's Problem as a Decision-Theoretic Problem*

		State of Nature		
		SB correct	**SB incorrect**	
You	**Take only what is in B2**	$1,000,000	$0	No dominant strategy
	Take what is in both boxes	$1,000	$1,001,000	

SB is incorrect, take what is in both boxes), neither of your two actions dominates the other.

Since you do not have a dominant strategy in the decision-theoretic representation of Fig. 3.2, there no longer exists a conflict between the expected-utility principle and the dominance principle. Now the sole determinant of whether you should take only what is in B2 or take what is in both boxes to maximize your expected utility are the probabilities that you associate with each state of nature. If the probability that SB is correct is greater than 0.5005, then you should take only what is in B2; if this probability is less than 0.5005, then you should take what is in both boxes; and if this probability is exactly 0.5005, then the expected utilities of your two courses of action are equal, and you would be indifferent between them.

How persuasive is this resolution of Newcomb's problem? If you believe that SB has no control over which state of nature occurs in Fig. 3.2, then SB is not properly a player in a two-person game, as erroneously assumed in Fig. 3.1; hence, the appropriate representation of Newcomb's problem is decision-theoretic. To be sure, the probabilities of being in each state are not specified by Newcomb's problem, so the decision-theoretic representation does not answer the question of whether you should take only what is in B2 or take what is in both boxes. However, this representation does demonstrate that there is no conflict between the dominance principle and the expected-utility principle.

On the other hand, if you believe that SB has some control over which state of nature occurs—which is a question quite different from whether he can predict your choice (which he almost surely can)—then he is not an entirely passive "state of nature," at least with respect to being correct. Hence, the game-theoretic representation of Fig. 3.1 is the appropriate one, though it is incomplete if no preferences for the outcomes are assumed of SB.

However, it must be said that there is nothing in the original statement of Newcomb's problem to indicate that SB's choices are anything but mechanistic—that is, the correctness of his prediction about your action is *not* assumed to depend in any way on your choice. Or, to put it another way, though you are assumed to exercise free will with respect to the action you take, SB exercises no free will with respect to what he puts in B2; his "choice" is dictated solely by his prediction.

The fact that SB's prediction is assumed to be almost surely correct would seem to imply that you are indeed playing a game against nature whose two states—SB correct or SB incorrect—occur with the same relative frequency, whatever you do. Given that this is the proper interpretation of Newcomb's problem, then Ferejohn's ingenious decision-theoretic reformulation of the problem convincingly resolves the presumed conflict between the dominance and expected-utility principles.

3.5. *Newcomb's Problem Symmetricized: Prisoners' Dilemma*

If one can dispose of Newcomb's problem in the above manner, it is still intriguing to ask what consequences the predictive ability assumed on the part of SB would have if *both* actors in Newcomb's problem could make genuine choices as players in a game. Moreover, if SB not only can predict but also can obtain a payoff, and P, like SB, can predict his adversary's choices, how does the game change?

To generalize the payoff matrix of Newcomb's problem, assume that the payoffs given in the Fig. 3.3 matrix are *utilities* of the outcomes to the row player (A) (not just ranks), A_4 being his best payoff, A_3 next best, and so on. I shall consider payoffs for the column player (B) shortly, using "A" and "B" instead of "P" and "SB" to emphasize the symmetry of the two players' roles.

The dominance principle says that player A should choose strategy a_2; the expected-utility principle says that player A should choose strategy a_1, given that A considers B's ability to predict his (A's) choices to be "sufficiently good." More precisely, if p is the subjective probability that A believes B's prediction about his strategy choice will be correct, then the expected-utility principle would prescribe that A should choose strategy a_1 if

$$A_3p + A_1(1 - p) > A_4(1 - p) + A_2p.$$

Figure 3.3 *Generalized Payoff Matrix for Player A in Newcomb's Problem*

		B		
		Predicts a_1	**Predicts a_2**	
A	a_1	A_3	A_1	
	a_2	A_4	A_2	← Dominant strategy

Key: $A_4 > A_3 > A_2 > A_1$, where A_i is the payoff (utility) to player A of an outcome

In Newcomb's problem, an asymmetry is assumed in both the abilities and actions of the two players in the prediction-choice game. SB (player B in Fig. 3.3) is assumed to be a phenomenally good guesser, but no such superior intelligence is attributed to the chooser (player A in Fig. 3.3). Furthermore, player B is assumed to make the first move, but in fact this gives him neither an advantage nor a disadvantage because his choice of what to put in the boxes (based on his prediction) is not communicated to player A. Thus, one could just as well assume that the two players make simultaneous choices; the essential nature of the game would remain unchanged.

The game does change, however, if not only player B can make predictions about player A's choices, but A can make predictions about B's choices as well. If player B's ranking of the outcomes duplicates player A's in Fig. 3.3—but now, with the rows and columns interchanged, A is assumed to be the predictor and B the chooser—the payoff matrix for player B will appear as in Fig. 3.4, with B_4 representing his best payoff, B_3 his next best, and so on. (If B was SB earlier, now think of A, as P, being able—like the biblical prophet—to make predictions about SB as well; the implications of mutual predictability will be developed below.) As was true for player A in the Fig. 3.3 game, here the dominance principle and the expected-utility principle prescribe different strategy choices for player B if he (B) considers player A's ability to predict his choices to be sufficiently good.

If the payoffs in the two asymmetrical prediction-choice games are combined into a single payoff matrix, the resulting game will be that depicted in Fig. 3.5 (in which only the players' strategies, but not their predictions about the other player's strat-

Figure 3.4   *Payoff Matrix for Player B*

		B	
		b_1	b_2
	Predicts b_1	B_3	B_4
A			
	Predicts b_2	B_1	B_2
			↑
			Dominant strategy

Key: $B_4 > B_3 > B_2 > B_1$, where B_j is the payoff (utility) to player B of an outcome

Figure 3.5   *Combined Payoff Matrix for Players A and B*

Player B

		b_1	b_2
Player A	a_1	(A_3,B_3)	(A_1,B_4)
	a_2	(A_4,B_1)	(A_2,B_2) ← Dominant strategy

↑
Dominant strategy

Key: $A_4 > A_3 > A_2 > A_1$, where A_i is the payoff (utility) to player A of an outcome

$B_4 > B_3 > B_2 > B_1$, where B_j is the payoff (utility) to player B of an outcome

Circled outcome is a Nash equilibrium

egy choices, are shown). The payoff matrix for this game gives the outcomes for both players, where, for each cell entry (A_i,B_j), A_i represents the payoff to the row player and B_j the payoff to the column player. Lo and behold, the ranking of outcomes by both players in this game generated by the symmetric play of Newcomb's prediction-choice game defines the classic 2×2 Prisoners' Dilemma game (see Section 2.4)!

Recall that the dilemma for the players in this game lies in the fact that whereas they both prefer payoff (A_3,B_3) to payoff (A_2,B_2), the former payoff is not a Nash equilibrium: each player has an incentive to shift to his second (dominant) strategy, given that the other player sticks to his first (dominated) strategy. If the other player chooses his second (dominant) strategy, one's own dominant strategy is again better. Both players are therefore motivated to "play it safe" and choose their dominant second strategies (a_2 and b_2), which—unfortunately for them—yields the "noncooperative" outcome (A_2,B_2) that both find inferior to the "cooperative" outcome (A_3,B_3).

3.6. A Solution to Prisoners' Dilemma

The fact that the problems of choice in Newcomb's problem and Prisoners' Dilemma are related should not obscure the fact that the latter is a two-person game—in which both players can make free and independent choices—whereas the former seems best conceptualized as a (one-person) game against nature, or a situation of decision making under uncertainty. However, the condition in the symmetric version of Newcomb's problem that each player knows that the other player can predict—with a high degree of accuracy—which strategy he will choose has a surprising consequence for the play of Prisoners' Dilemma: it provides an incentive for each player *not* to choose his second dominant strategy (a_2 or b_2).

True, if one player knows that the other player will almost surely choose his second strategy, then he should also choose his second strategy to insure against receiving his worst payoff (A_1 or B_1). As a consequence of these choices, the noncooperative payoff (A_2 or B_2) will be chosen.

But now assume that one player knows that the other player plans—at least initially—to select his first strategy. Then one would ordinarily say that he should exploit this information and choose his second strategy, thereby realizing his best payoff (A_4 or B_4). But this tactic will not work, given the mutual predictability of choices I assumed on the part of both players in this symmetric version of Newcomb's problem. For any thoughts by one player of "defecting" from his strategy associated with the cooperative but unstable payoff, (A_3, B_3), would almost surely be detected by the other player. The other player then could exact retribution—and at the same time prevent his worst outcome from being chosen—by switching to his own noncooperative strategy. Thus, the mutual predictability of strategy choices assumed in the symmetric version of Newcomb's problem helps to insure against noncooperative choices by *both* players and stabilize the cooperative solution to Prisoners' Dilemma.

More formally, assume player A contemplates choosing either strategy a_1 or a_2 and knows that player B can correctly predict his choice with probability p and incorrectly predict his choice with probability $1 - p$. Similarly, assume that player B, facing the choice between strategy b_1 and b_2, knows that player A can correctly predict his choice with probability q and incorrectly predict his choice with probability $1 - q$. Given these probabilities, I shall show that there exists a "choice rule" that *either* player can adopt that will induce the other player to choose his cooperative strategy—based on the expected-utility criterion—given that the probabilities of correct prediction are sufficiently high.

A *choice rule* is a conditional strategy based on one's prediction of the strategy choice of the other player. In the calculation to be given shortly, I assume that one player adopts a choice rule of *conditional cooperation:* he will cooperate (i.e., choose his first strategy) if he predicts that the other player will also cooperate by choosing his first strategy; otherwise, he will choose his second (noncooperative) strategy.

Assume player B adopts a choice rule of conditional cooperation. Then if player A chooses strategy a_1, B will correctly predict this choice with probability p and hence will chose strategy b_1 with probability p and strategy b_2 with probability $1 - p$. Thus, given conditional cooperation on the part of B, A's expected utility from choosing strategy a_1, $E(a_1)$, will be

$$E(a_1) = A_3p + A_1(1 - p).$$

Similarly, his expected utility from choosing strategy a_2, $E(a_2)$, will be

$$E(a_2) = A_4(1 - p) + A_2p.$$

$E(a_1)$ will be greater than $E(a_2)$ if,

$$A_3p + A_1(1 - p) > A_4(1 - p) + A_2p,$$
$$(A_3 - A_2)p > (A_4 - A_1)(1 - p),$$
$$\frac{p}{1 - p} > \frac{A_4 - A_1}{A_3 - A_2}.$$

It is apparent that the last inequality is satisfied, and $E(a_1) > E(a_2)$, whenever p (in comparison to $1 - p$) is sufficiently large, i.e., whenever p is sufficiently close to 1. If, for example, the utilities associated with player A's payoffs are $A_4 = 4$, $A_3 = 3$, $A_2 = 2$, and $A_1 = 1$, then the expected utility of player A's first strategy will be greater than that of his second strategy if

$$\frac{p}{1 - p} > \frac{4 - 1}{3 - 2},$$
$$p > 3(1 - p),$$
$$4p > 3,$$
$$p > \tfrac{3}{4}.$$

That is, by the expected-utility criterion, player A should choose his first (cooperative) strategy if he believes that player B can correctly predict his strategy choice with a probability greater than ¾, given that player B responds in a conditionally cooperative manner to his predictions about A's choices. Note that whatever the utilities consistent with player A's ranking of the four outcomes are, p *must* exceed ½ because $(A_4 - A_1) > (A_3 - A_2)$.

What happens if player B adopts a less benevolent choice rule? Assume, for example, that he always chooses strategy b_2, whatever he predicts about the strategy choice of player A. In this case, if A now adopts a conditionally cooperative choice rule, he will choose strategy a_1 with probability $1 - q$ and strategy a_2 with probability q. By the symmetry of the game, if the roles of A and B are reversed, one can show, in a manner analogous to the comparison of the expected utilities of strategies given previously for player A, that $E(b_1) > E(b_2)$ if

$$\frac{q}{1 - q} > \frac{B_4 - B_1}{B_3 - B_2},$$

or whenever q (in comparison to $1 - q$) is sufficiently large, i.e., whenever q is sufficiently close to 1. Subject to this condition, therefore, player B would *not* be well advised always to choose strategy b_2 if player A adopts a conditionally cooperative choice rule. Clearly, if both the previous inequalities are satisfied, A and B each do better choosing their cooperative strategies, a_1 and b_1, respectively, to maximize their expected utilities, given that each player follows a choice rule of conditional cooperation.

3.7. Cooperation or Noncooperation?

So far I have shown that if one player—call him the *leader*—(i) adopts a conditionally cooperative choice rule and (ii) can predict the other player's strategy choice with a sufficiently high probability, the other player—call him the *follower*—maximizes his own expected utility by cooperating also, given that he can detect lies on the part of the leader with a sufficiently high probability. Thereby both players "lock into" the cooperative solution, which—it will be remembered—is unstable in Prisoners' Dilemma when the players do not have the ability to predict each other's strategy choices (see Section 2.4).

There is one question that remains, however. Given that the follower maximizes his expected utility by cooperating when the leader adopts a choice rule of conditional cooperation, how does the follower know when the leader adopts such a choice rule in the first place? The answer is that he does not (a follower can predict a choice but not a choice rule) unless the leader announces his intention to adopt this choice rule.

To escape the dilemma, therefore, one must assume that there is some communication between the players—specifically, that one player (the leader) announces a choice rule to which the other player (the follower) responds. If neither player takes the initiative, nothing can happen; if both players take the initiative simultaneously and announce the choice rule of conditional cooperation, each presumably will await a commitment from the other before committing himself, and again nothing will happen. Should the players simultaneously announce different choice rules, the resulting inconsistencies may lead to confusion, or possibly an attempt to align the rules or distinguish the roles of leader and follower.[12]

The only clean escape from the dilemma, therefore, occurs when the two players can communicate and take on the distinct roles of leader and follower. Although, strictly speaking, permitting communication turns Prisoners' Dilemma into a game that is no longer wholly noncooperative, communication alone is not sufficient to resolve the dilemma without mutual predictability. For what is to prevent the leader from lying about his announced intention to cooperate conditionally? And what is to prevent the follower from lying about his announced response to select his cooperative strategy?

The insurance against lies that players have with mutual predictability is that the lies can be detected with probabilities p and q. If these probabilities satisfy the previous inequalities, then it pays for the follower to cooperate in the face of a choice rule of conditional cooperation, and the leader to cooperate by then choosing his cooperative strategy, too. Otherwise, the insurance both players have against lying will not be sufficient to make cooperation worth their while, and they should, instead, choose their noncooperative dominant strategies. I conclude, therefore, that a mutual ability to predict choices on the part of both players offers them a mutual incentive to choose their cooperative strategies.

It is worth noting that the solution to Prisoners' Dilemma proposed here has some similarities to the solution of this game prescribed by "metagame theory," but there are also some significant differences.[13] In this theory, the successive iteration of conditional

---

12. The so-called Stackelberg solution in duopoly theory in economics also distinguishes between a "leader" and a "follower." See John M. Henderson and Richard E. Quandt, *Microeconomic Theory: A Mathematical Approach*, 2nd Ed. (New York: McGraw-Hill, 1971), pp. 229–231.

13. Nigel Howard, *Paradoxes of Rationality: Theory of Metagames and Political Behavior* (Cambridge, MA: MIT Press, 1971). For an overview of this theory, with examples, and some of the controversy it has generated, see Brams, *Paradoxes in Politics*, Chaps. 4 and 5. For a refinement of Howard's notion of metarational outcomes, see Niall M. Fraser and Keith W. Hipel, Solving complex conflicts, *IEEE Trans. Systems, Man, and Cybernetics* SCM-9, 12 (December 1979), 805–816.

strategies by the players yields some that render the cooperative outcome in equilibrium for the players.

The choice rule of conditional cooperation I have posited assumes, in effect, the existence of a first-level (or "leader") metagame, which gives the follower a motive to cooperate in response to the leader's tit-for-tat conditional strategy. But unlike Howard, I do not carry the analysis to a second-level (or "follower-leader") metagame in which the leader is given a motive to play tit-for-tat against the follower's own tit-for-tat policy, once removed.

The reason I eschew this stepwise backward reasoning is that it seems unnecessary if—as earlier assumed of SB in Newcomb's problem—players' predictions (in the *preplay* leader-follower negotiation phase of the game) precede their choices (in the *play* of the game). Clearly, the proposal of conditional cooperation by the leader in the preplay phase is sufficient to initiate the process of cooperation. Then, however the players become aware of each other's powers of prediction, prediction probabilities that satisfy the previous inequalities are sufficient to protect the players against either's reneging on an agreement. For given that each player knows that the other player's probability of predicting his own strategy choice is sufficiently high, he knows that he probably cannot get away with a sudden switch in his strategy choice in the play of the game, because this move will already have been anticipated with a high probability in the preplay phase. Hence, the assumption that (preplay) predictions precede (play-of-the-game) choices—and both players know this—deters "last-minute" chicanery that would render the cooperative outcome unstable.

The advantage offered by a leader-follower model that distinguishes unambiguously between the preplay and play phases of a game lies not only in its ability to truncate the iterative calculations of metagame theory; it also offers an advantage in highlighting the circumstances under which players would come to harbor tit-for-tat expectations in the first place. If they come to realize, in the preplay phase of the game, that their later choices in the play of the game are, to a sufficiently high degree, predictable, they will be purged of their incentive to violate an agreement, given that they are expected-utility maximizers.

In this manner, the leader-follower model suggests circumstances under which an *absolutely* enforceable contract will be unnecessary. When the prediction probabilities of the players are sufficiently high (which depends on the utilities assigned by the players to the outcomes), an agreement to cooperate—reached in leader-follower negotiations in the preplay phase of the game—

can be rendered "enforceable enough" so as to create a probabilistic kind of equilibrium that stabilizes the cooperative outcome.

By introducing probabilities of correct prediction *as parameters* in the preplay phase of a game, one is able to place the metagame solution to Prisoners' Dilemma within a rational-choice framework. What emerges as a solution is, in essence, a *consequence* of the rationality assumption (i.e., that players maximize expected utility) rather than the *assumption* that there exists some kind of consciousness of predictability among players. This is not to denigrate the metagame solution, but rather to show that there is a compelling rationale for its existence within a rational-choice framework.[14]

3.8. Implications of Mutual Predictability for SB and P

Do the leader-follower and mutual predictability assumptions make any sense in a theological context descriptive of games that might be played between SB and P? In the Bible, the covenant that God established with the Israelites seems to have some of the earmarks of a relationship of predictability and trust, beneficial to both parties, with God at the helm. In fact, His leadership was not just spiritual but occasionally very down-to-earth, as when He, through Moses, led the Israelites out of Egypt and then inundated Pharaoh's chariots in the Red Sea.

If the biblical covenant was more than a legal contract, it was also less, because it was not something legally binding and enforceable as such. It depended on good will, backed up, of course, by the possibility of reprisal if the human players did not live up to its (often unwritten) terms.

To the degree that there is mutual predictability, the need for threats and reprisals diminishes, even in games as taxing for the players as Prisoners' Dilemma. Binding and enforceable contracts

---

14. This framework has been extended in Steven J. Brams, Morton D. Davis, and Philip D. Straffin, Jr., The geometry of the arms race, *Int. Studies Quarterly* 23, 4 (December 1979), 567–588; see the comment on this article by Raymond Dacey, Detection and disarmament, pp. 589–598, and Brams, Davis, and Straffin, A reply to "Detection and disarmament," pp. 599–600, in the same issue. Further refinements in this framework can be found in Dacey, Detection, inference and the arms race, in *Reason and Decision*, Bowling Green Studies in Applied Philosophy, Vol. III-1981, ed. Michael Bradie and Kenneth Sayre (Bowling Green, OH: Applied Philosophy Program, Bowling Green State University, 1982), pp. 87–100.

also become unnecessary, for violations will be predictable with a high probability in the preplay phase of the game, making possible the application of appropriate sanctions to the violator in the play of the game. But because such retribution works to the disadvantage of both players, the ability of both players to predict each other's choices also serves to reinforce trustworthy behavior, which is exactly what is not encouraged in Prisoners' Dilemma without mutual predictability.

In the Prisoners' Dilemma in Fig. 2.3 (fourth Knowability Game), assume that SB can predict P's expectations about knowability/unknowability. Thereby he can try to induce the cooperative solution by announcing: "I'll be knowable if you expect me to be unknowable; otherwise, I won't." This, perhaps, sounds bizarre: "I'll be knowable on condition that you *not* expect it—and so not investigate me."

Yet, for reasons given in Section 2.3, I think this attitude is eminently sensible. Just as the biblical God often appeared ambivalent, it is not unreasonable to suppose that a SB would desire that nothing be known or expected of him. This way he can conjure up, by his own words and actions, the image he wants to project, without worrying about an inquisitive P who thinks his investigation of SB will be fruitful.

When SB is successful in this endeavor, he can transform the Pareto-inferior (2,2) outcome in the fourth Knowability Game into the Pareto-superior (3,3) outcome. This outcome, of course, is better for P as well; moreover, P has good reason to cooperate by eschewing investigation, because if he does not, and SB predicts that P will investigate him, SB will defect himself—that is, be unknowable. Then both players will have their primary goals thwarted, whereas (3,3) allows both to achieve these goals: for SB—that P does not investigate him; for P—that he confirms SB's knowability, or, as is the case at (3,3), disconfirms his unknowability.

"Disconfirmed" is probably a misnomer here—at least as far as SB is concerned—because what he (SB) really has an opportunity to do, if (3,3) is chosen, is write on a clean slate, so to speak. That is, he can be knowable *on his own terms*, by virtue of the fact that P decides not to investigate him. Whether P is pleasantly or unpleasantly surprised, he does gain insight into SB, which is, by assumption in this game, his highest aim.

If SB, as the presumed leader, induces a cooperative response from P as the follower in the fourth Knowability Game because of his predictive power, what assurance does P have that SB will act cooperatively, too? After all, SB could do still better in the play phase by appearing unknowable, which would confirm SB's un-knowability (best for SB, worst for P).

The problem with this kind of deception on the part of SB, which I shall explore further in Chapter 6, is that it might work in any single play of Prisoners' Dilemma but would undoubtedly undercut P's trust of SB in future play. Because it is in SB's as well as P's interest to lock into (3,3) rather than (2,2), SB would be well advised not to tarnish his reputation for cooperation in Prisoners' Dilemma by too many double-crosses in this game.

In other words, it is to SB's advantage to keep his word, by sticking to the choice rule of conditional cooperation, precisely to try to steer both players clear of the Pareto-inferior outcome. Thus, if only SB has omniscience, or, even less-than-perfect pre-dictive power, and is a continuing player, he should indicate by his actions that he will be (predictably) cooperative if P is—accord-ing to SB's best estimate—in order that both players can attain (3,3) in Prisoners' Dilemma.

But predictability, and a choice rule of conditional coopera-tion, is not just a device SB can use to induce a mutually beneficial outcome in Prisoners' Dilemma. Consider the Revelation Game (Fig. 2.1) and the effects of SB's announcement—or *some* indication if an announcement would give his presence away—that he will reveal himself if he predicts P will believe in his existence, other-wise he will not. If q is SB's probability of correctly predicting P's strategy choice, and P's utilities are consistent with his ranks in this game (A_4 utility for P's best outcome, A_3 for his next best, etc.), his expected utility from believing is

$$E(B) = A_4q + A_2(1 - q),$$

from not believing,

$$E(\overline{B}) = A_1(1 - q) + A_3q.$$

Now

$$E(B) - E(\overline{B}) = (A_4 - A_3)q + (A_2 - A_1)(1 - q) > 0$$

for $0 \le q \le 1$, since $A_4 > A_3 > A_2 > A_1$. That is, whatever value SB's prediction probability, q, assumes, P maximizes his expected utility by believing in SB's existence because $E(B) > E(\overline{B})$. Hence, assume that P always chooses B.

Then SB will reveal himself with probability q, leading to outcome (3,4) in Fig. 2.1, and will not reveal himself with probability $1 - q$, leading to outcome (4,2). But note that whatever q is, P's strategy of believing (B) is *always* optimal, given that SB follows the choice rule of conditional cooperation.

This is true despite the fact that B is not a dominant strategy for P. In effect, SB's choice rule of conditional cooperation makes it dominant, at least based on the expected-utility criterion. Furthermore, it is easy to show that if $q > (A_3 - A_2)/(A_4 - A_2)$, $E(B) > A_3$, yielding a higher expected payoff for P than that which he obtains when the Pareto-inferior Nash equilibrium, (2,3) in Fig. 2.1, is chosen by the players.

Similarly, SB benefits from this choice rule. Given that P chooses his (optimal) strategy B with certainty, SB's expected payoff from conditioned cooperation is

$$E(SB) = B_3q + B_4(1 - q).$$

Since $B_4 > B_3$, $E(SB) > B_3$ except when $q = 1$ [which gives $E(SB) = B_3$]. Thus, SB's choice rule of conditional cooperation will, except at q = 1 when there is equality, give him a higher expected payoff than he would obtain even from the Pareto-superior (3,4) outcome in Fig. 2.1, at which he obtains 3 (B_3). It is therefore to SB's advantage to be a good, but not a perfect, predictor!

All in all, then, conditional cooperation resoundingly benefits both players in the Revelation Game, but in an unexpected way. P *always* has an incentive to believe, whatever q is; if $q > (A_3 - A_2)/(A_4 - A_2)$, he will do even better, on the average, than what the Pareto-inferior (2,3) Nash equilibrium in Fig. 2.1 would give him (3). SB will do better, too; to wit, given that P always chooses B and $q \neq 1$, SB will obtain a greater expected payoff than even the Pareto-superior (3,4) outcome would give him (3).

These seemingly idyllic results of conditional cooperation in the Revelation Game do, however, have a dark side. First, whatever the value of q, SB will have an incentive always to choose nonrevelation, given that P chooses to believe, to induce (4,2), his best outcome. This, in turn, will induce P not to believe, leading back to the Nash equilibrium (2,3). Thus, P's inability to hold SB to conditional cooperation—because he is unable to predict in the preplay phase of the game that SB will indeed reveal himself if P believes—renders the cooperative choices/rules unstable.

Prediction probabilities are more consequential in the fourth Knowability Game (Prisoners' Dilemma), as I showed, but here, too, P's inability to check in the preplay phase that SB will adhere to the choice rule of conditional cooperation could lead to P's exploitation. I suggested, though, that SB, as a continuing player, might have good reason not to double-cross P in games, like this Prisoners' Dilemma, played repeatedly. I shall return to the assumption of repeated play of such games in Chapter 6.

3.9. Conclusions

The ability to predict with certainty another player's strategy choice in a game, particularly when he does not have a dominant strategy, seems a natural one to attribute to a superior being. Yet, SB's omniscience so defined provides no escape from the Pareto-inferior outcome in the Revelation Game (Fig. 2.1). Had P possessed this ability, though, it would have induced SB, aware of it, to reveal himself and P to believe in his existence. But endowing P with omniscience seems far-fetched if not preposterous.

If SB is assumed to be only partially omniscient, it is reasonable to attribute to him a certain probability of making a correct prediction, which is done in Newcomb's problem without being precise about the probability value. The problem for P in this situation is the apparent contradiction between the dominance principle and the expected-utility principle. Following Ferejohn, however, I showed that the conflict between these two principles can be persuasively resolved if Newcomb's problem is reformulated as a decision-theoretic problem rather than as a game.

In the decision-theoretic representation, which seems accurately to reflect the original statement of the problem, neither action is dominant for the person making the choice of which box(es) to choose. In the absence of a dominant strategy, therefore, the expected-utility principle cannot run amok of the dominance principle.

To divest Newcomb's problem of the asymmetry in predictive capabilities and choices assumed of SB and P, I next supposed that both players know that the other player can predict their choices with a high degree of accuracy. Combining into a single game the two games generated by both sets of choices, I showed that choices in the resulting game defined a Prisoners' Dilemma (Fig. 3.5).

Applied to the fourth Knowability Game (Fig. 2.3), the additional condition that P and SB know that each's strategy choices are almost surely predictable by the other player gives each an incentive to choose his cooperative—rather than noncooperative but dominant—strategy, given that one player (say, SB as the leader) adopts a choice rule of conditional cooperation to which the other player (P as the follower) responds. This resolution of the dilemma, while consistent with the solution offered by meta-game theory, rationalizes this solution: instead of assuming that players can successively predict each other's strategy choices before the game is played, it follows from calculations that maximize each player's expected utility.

Whereas the probabilities of correct prediction may be consequential in inducing the cooperative outcome in the fourth Knowability Game, they have no effect on inducing P to believe in the Revelation Game (though they do affect the players' expected payoffs in this game). Whatever SB's probability of prediction, it is rational for P to believe in SB's existence, given that SB is somehow able to indicate conditional cooperation on his part (revelation if P believes, nonrevelation otherwise); SB will also benefit from following this choice rule.

SB, however, has an incentive to renege on conditional cooperation to do even better, which presumably will be tempting if P cannot hold SB to the choice rule by his own predictive ability. In Chapter 5, I shall pursue this line of reasoning in greater depth, showing how SB's omnipotence can be used to force his best outcome in the Revelation Game.

In the present chapter, my main purpose was to introduce some potentially beneficial effects of SB's omniscience and partial omniscience. Next, in Chapter 4, I will examine some ill effects of omniscience and consider how SB might ameliorate them.

FOUR

The Paradox of Omniscience and the Theory of Moves

4.1. Introduction

In Chapter 3 I indicated how one can calculate, from players' payoffs, probabilities that indicate thresholds at which the cooperative solution in Prisoners' Dilemma can be rendered stable and the dilemma thereby circumvented. In the fourth Knowability Game, this resolution depended on the predictive abilities of P as well as SB, which it may be unreasonable to assume. On the other hand, in the less ethereal world of international politics, the presumption that the superpower arms race is a Prisoners' Dilemma, and that both sides have these predictive abilities—based on their intelligence capabilities, supported by reconnaissance satellites and other detection equipment—seems quite reasonable.[1]

---

1. See citations in note 14, Chapter 3.

Still, one-sided omniscience and partial omniscience have consequences, and in this chapter I shall further explore both the opportunities and difficulties that these abilities create for SB. In the Bible, God's clairvoyance often seems less than total, though He does prefigure certain events, such as Jacob's conflict with Esau and Samson's deliverance of Israel from the Philistines, at the prenatal stage. Also, he forewarns Cain of sin "at the door; its urge is toward you" (Gen. 4:7).

Yet, it is not clear that God anticipated Cain's murder of Abel, or Adam and Eve's earlier transgressions, from the questions (perhaps rhetorical) that He asked of them after they had sinned. More to the point, however, God is often frustrated when events that he seems not to have anticipated turn against Him, as when the people of Israel demand a king and Saul is subsequently anointed.

I defined omniscience in Section 3.2, showing that for SB it had no effect on the (Pareto-inferior) outcome he was able to implement in the Revelation Game. In Section 4.2, I shall demonstrate for a specific game how this definition can lead to a paradox in its play and suggest three ways in which this paradox may be resolved, only one of which is satisfactory.

Next, I shall present a procedure that, through the transformation of an outcome matrix, leads to a resolution of the paradox. This analysis is based on what I call the "theory of moves," which I shall develop further in subsequent chapters when I define different notions of omnipotence and immortality. The theory postulates dynamic rules of play that allow for sequential moves and countermoves by the players after they have made their initial strategy choices. As will be demonstrated, it eliminates some of the problems that omniscience may create for SB.

A variation on the paradox of omniscience, however, underscores the fact that these rules are not always sufficient to induce a stable Pareto-superior outcome. This problem is illustrated by what I call the Testing Game, in which SB's omniscience forces P to prepare himself for a test that may not be rational for SB to administer, thereby hurting both players.

By contrast, in games not vulnerable to a paradox of omniscience, such as Prisoner's Dilemma, the new rules can lead to long-term stable outcomes that do not necessarily coincide with Pareto-inferior Nash equilibria. I call these new equilibria, which provide the cornerstone of the theory of moves that I shall sketch, "nonmyopic equilibria," and will illustrate them in different games.

4.2. The Paradox of Omniscience[2]

The paradox of omniscience is illustrated by the following two-person game played between P and SB, each of whom has two strategies, "compromise" (C) and "don't compromise" ($\overline{C}$), as shown in Fig. 4.1. The choices of these strategies by the players, who are assumed to be on a "collision course," have the following consequences:

1. If both do not compromise, the result will be disastrous (the worst outcome for both).
2. If both compromise, they will reach a satisfactory agreement (the next-best outcome for both).

2. This section and the next are based largely on Steven J. Brams, Mathematics and theology: game-theoretic implications of God's omniscience, *Math. Mag.* 53, 5 (November 1980), 277–282. This article provoked a response from Ian Richards on the "responsible use of mathematics" in News and Letters, *Math. Mag.* 54, 3 (May 1981), 148; I replied in *Math. Mag.* 54, 4 (September 1981), 219.

Figure 4.1   *Outcome Matrix of Chicken*

		P	
		Compromise (C)	**Don't** **compromise** ($\overline{C}$)
SB	**Compromise** (C)	Cooperation (3,3)	P prevails over SB (2,4)
	Don't **Compromise** ($\overline{C}$)	SB prevails over P (4,2)	Disaster (1,1)

↑
Optimal
strategy of P
if he is aware
that SB is
omniscient

Key: 4 = best; 3 = next best; 2 = next worst; 1 = worst
Circled outcomes are Nash equilibria

3. If one player compromises and the other does not, the one who does not prevails (best outcome) over the one who does (next-worst outcome).

Clearly, this is a game in which both players face disaster (1) if they do not compromise, whereas both do relatively well (3) if they do. However, each player can obtain his best outcome (4) only if he does not compromise and the other player does, which hurts the player who does compromise but is not as disastrous for him (2) as the case in which both do not compromise and head for collision.

Neither player has a dominant strategy in this game: C is better if the other player chooses $\overline{C}$, but $\overline{C}$ is better if the other player chooses C. Nevertheless, though one player's preferred strategy depends on the choice of the other player, this game has two Nash equilibria, (4,2) and (2,4), from which neither player has an incentive to depart unilaterally because he will do worse if he does [at either (1,1) if, say, P departs from (4,2), or (3,3) if SB departs from (4,2)]. The problem, of course, is that one of these equilibria is better for P, the other for SB, so in the absence of other assumptions it is hard to say which outcome, if either, rational players would choose.

The compromise (3,3) outcome, as in Prisoners' Dilemma (Fig. 2.3), is not in equilibrium. However, though the Fig. 4.1 game has some similarities to Prisoners' Dilemma (the next-worst and worst outcomes are interchanged in these two games), this game confronts the players with radically different problems. Principally, the two Nash equilibria in the Fig. 4.1 game are in competition, whereas the one Nash equilibrium in Prisoners' Dilemma is Pareto-inferior.

In the game-theoretic literature, this game is called "Chicken." I have discussed its implications, particularly in politics, elsewhere.[3] This is a game that has been used to model, among other situations, the Cuban missile crisis of October 1962 in which the United States and the Soviet Union faced each other in a nuclear confrontation.[4]

In the Bible, there are a number of instances in which such

3. Steven J. Brams, *Game Theory and Politics* (New York: Free Press, 1975), Chap. 1; and Brams, *Paradoxes in Politics: An Introduction to the Nonobvious in Political Science* (New York: Free Press, 1976), Chap. 5.

4. See Steven J. Brams, Deception in 2 $\times$ 2 games, *J. Peace Sci.* 2 (Spring 1977), 177–203, where I argue that a game other than Chicken better models the Cuban missile crisis.

figures as Adam and Eve, Cain, Saul, and even Moses confronted God and, in addition, were forewarned or had forebodings that the consequences of their defiance would not be benign. Moreover, in these conflicts God did not always prevail. For example, after Moses's intercession, He reversed His original position to destroy the Israelites, following their idolatry at Mount Sinai, and accepted a kind of compromise solution in which only some were killed. Indeed, although one might be hard put to say that disaster befell God after Adam and Eve sinned, Cain murdered his brother, Abel, or Saul disobeyed God's mouthpiece, the prophet Samuel, He certainly did not appear elated by these violations of His commands and precepts.

My point is that while the game of Chicken obviously does not mirror all conflicts and confrontations that man has had or might have with God, it is not implausible to think of man and God as being occasionally on a collision course, with possibly doleful results for both players. As in most games that the God of the Old Testament played with man, Chicken is a game of partial conflict in which both players can, relatively speaking, "win" [at (3,3)] or "lose" [at (1,1)] simultaneously.

From Section 3.2 recall that omniscience is the ability of one player to predict the strategy choice of the other player. As before, I assume that SB has this ability, P is aware that SB possesses it, and both players seek to achieve their best possible outcome while cognizant of each's ability/awareness.

Specifically, in the game of Chicken, if P is aware of SB's omniscience, he should not compromise, because he knows that SB will predict this choice on his part. Thereby SB will be forced to compromise himself to prevent his worst outcome (1) from occurring and to salvage instead his next-worst outcome (2).

Thus, if P is aware of SB's omniscience, he would prevail over him, yielding the Nash equilibrium (2,4) in Chicken. The fact that SB's superior predictive ability, and P's awareness of it, help P and hurt SB I call the *paradox of omniscience*. This is a paradox, I believe, because one would not expect this superior ability of SB to impede his position—the outcome he can ensure—in a game. Yet, it is precisely his omniscience, and P's awareness of it, which ensures that P obtains his best outcome and SB does not.[5]

5. For other examples in which additional information may hurt, see Y. C. Ho and I. Blau, A simple example on informativeness and competitiveness, *J. Optimization Theory Appl.* 11, 4 (April 1973), 437–440; Ariel Rubinstein, A Note on the duty of disclosure, *Economic Lett.* 4 (1979), 7–11; Martin Shubik, *Game Theory in*

4.3.   A Resolution of the Paradox of Omniscience

In order to avoid misunderstanding, I want to stress that I have made only the usual assumptions about the play of the game I have described: both players (i) have complete information about the outcome matrix, (ii) make simultaneous choices, (iii) cannot communicate with each other, and (iv) are rational in the sense of seeking to achieve their best possible outcomes. The only assumptions I have added to these in the play of Chicken is that (v) SB is omniscient and (vi) P is aware of his omniscience.

To try to resolve the paradox of omniscience, I want first to relax assumption (iii) and then allow for the possibility of retaliation by SB or retaliation by both players in a sequence of moves and countermoves. I suggest that this reformulation of the game permits at least three possible ways for SB to counter the paradox:

I. Before play of the game, SB can threaten P: "If you compromise, I will; otherwise, I won't." This threat is not credible for two reasons: (i) if P compromises and SB predicts this, it is not in SB's interest to compromise, too, obtaining his next-best outcome (3) at CC; rather, SB should not compromise and obtain his best outcome (4) at $\overline{C}C$; (ii) if P does not compromise, and SB predicts this, SB has no alternative but to compromise to prevent his worst outcome (1) at $\overline{CC}$ from occurring. Therefore, P should not compromise to induce (2,4), his best outcome.

In the next two possible solutions, I assume that not only is communication permissible—making threats possible—before the play of the game, but also that one or both players can retaliate after the play of the game. In solution II below, I assume that only SB can retaliate by switching from his strategy of C to $\overline{C}$, or vice versa. In solution III, I assume that both players can retaliate by switching strategies in a strict sequence—first one, then the other, and so on—without the possibility of retracting a move once it is

the Social Sciences: Concepts and Solutions (Cambridge, MA: MIT Press, 1982), p. 274; and Richard Engelbrecht-Wiggans and Robert J. Weber, Notes on a sequential auction involving asymmetrically-informed bidders, Int. J. Game Theory (forthcoming). The paradox of omniscience, while unfortunate for SB, may be regarded as a price he pays for information about the future. Such foreknowledge, as Vladimir Lefebvre pointed out to me (personal communication, March 25, 1980), is of value itself, reaffirming the economist's adage that everything has a price. Or, to put it slightly differently, even SB must make trade-offs, though new rules of the game may save him, as I shall presently show.

made. In the latter solution, when one player has no interest in making a further move, the game terminates and its outcome occurs where play stops. Here are the two possible solutions, with an analysis of their merits:

II. Before play of the game, SB can threaten P: "Choose the strategy that you wish, but I reserve the right to retaliate by switching my strategy after the play of the game." This is again an incredible threat because, by virtue of his omniscience, SB can predict P's choice and make his own best response to it. Reserving the right to move after the play of the game gives SB no additional advantage.

III. Before play of the game, SB can threaten P: "Choose the strategy that you wish, but we each will have the right to move and countermove an indefinite number of times until one of us stops, whence the outcome reached is implemented." I shall now show that outcome (3,3) is the *only* outcome from which neither player will have a desire to move if it is chosen. Moreover, if any of the other three outcomes is chosen, moves and countermoves will terminate at an outcome inferior to (3,3) for at least one player, thereby establishing the choice of compromise by one, and then of necessity the other, as their only rational choice. Hence, (3,3) is engendered under the rules of solution III.

Assume (2,4) is chosen, which is the outcome in the original game and creates the paradox of omniscience. Then by moving the process to (1,1), SB would induce P to move it to (4,2), where the process would stop since it is best for SB. But this outcome is next worst for P.

Assume (4,2) is chosen. Then analogous moves by P and SB, respectively, would move the process to (2,4), which is next worst for SB.

Assume (1,1) is chosen. Then the player who moves first (P or SB) would get his next-worst outcome [(2,4) or (4,2)], and the other player would obtain his best outcome and have no incentive to countermove. In either event, or if no player moved initially, at least one player would obtain an outcome inferior to his next-best outcome of 3.

Now assume (3,3) is chosen. Then if SB initiates the move process, it will proceed as follows: (3,3) → (4,2) → (1,1) → (2,4), so the final outcome, (2,4), is inferior to (3,3) for SB. Similarly, if P initiates the first move, the process will proceed as follows: (3,3) → (2,4) → (1,1) → (4,2), so the final outcome, (4,2), is inferior to (3,3) for P. Hence, neither P nor SB will, if they both choose compromise, have an incentive to deviate initially from (3,3).

Because the other three outcomes will (eventually) result in the implementation of an outcome inferior to 3 for the player who does *not* choose compromise, each player will have an incentive to choose compromise. Since the resulting (3,3) outcome, given the rules I have specified in the third solution, is stable, rational players who desire to ensure their best possible outcome will choose it.

This solution needs to be slightly qualified. If both players choose not to compromise, resulting in (1,1), the player who can hold out longer, forcing his opponent to move first, obtains his best outcome of 4. Presumably this would be SB in most situations. But whoever it is, the other player has a foolproof counterstrategy—compromise initially. For if the recalcitrant player (say, SB) does not, his best outcome, (4,2), is transformed by one move and one countermove into (2,4), which is next worst for him. Thus, he should compromise, too.

More generally, as long as moves are strictly sequential and no backtracking is allowed, the unique stable (and desirable) outcome for both players is (3,3), regardless of who deviates first. Even if one player believes he can hold out longer at (1,1), if it is chosen initially, a common perception of this fact by both players will lead one—and hence both, for reasons given above—to choose compromise. Only if the perceptions of the two players differ— each thinks he can force the other to capitulate first when they start out at (1,1)—will compromise not be the rational choice of both players. But I assume this is not the case: the players' perceptions agree.

Thus, given the rules of solution III, outcome (3,3) is implemented. Paradoxically, perhaps, the possibility of moves and countermoves by both players—not just SB's ability to retaliate unilaterally—enables SB to counter the paradox his omniscience induces in the play of Chicken. Note that both the verbal threat of solution I, and the threat of possible retaliation after the play of the game of solution II, are insufficient to resolve the paradox. Only when reprisals by *both* players are permitted is SB able to wipe out the disadvantage that his omniscience encumbered him with in the original play of Chicken.

In Section 4.4, I shall formalize somewhat this informal analysis of Chicken. In addition, I shall show how it can be applied to other games—in particular, Prisoners' Dilemma. This solution provides a resolution of this dilemma in the fourth Knowability Game (Fig. 2.3) different from that given by the mutual predictability and leader-follower assumptions discussed in Sections 3.6– 3.8.

4.4. *Theory of Moves*[6]

In Chapter 1 I defined a game as the sum-total of the rules that describe it. In classical game theory, rule I below is usually the only rule of play assumed in normal-form or matrix games; the theory of moves postulates three additional rules of play that define a *sequential game:*

I. Both players simultaneously choose strategies, thereby defining an *initial outcome.*[7]

II. Once at an initial outcome, either player can unilaterally switch his strategy and change that outcome to a subsequent outcome in the row or column in which the initial outcome lies.

III. The second player can respond by unilaterally switching his strategy, thereby moving the game to a new outcome.

IV. The alternating responses continue until the player whose turn it is to move next chooses not to switch his strategy. When this happens, the game terminates, and the outcome reached is the *final outcome.*

Note that the sequence of moves and countermoves is *strictly alternating:* first, say, the row player moves, then the column player, and so on, until one stops, at which point the outcome reached is final.

How does a rational player determine whether he should move at a particular stage? I assume he performs a backward induction analysis, based on the *game tree* of possible moves that could be set off if he departed from the initial outcome. I shall develop this analysis in terms of row (R) and column (C) as the players since the *theory of moves* I shall somewhat informally describe, based on rules I–IV and other rules to be discussed later,

---

6. This section is based largely on Steven J. Brams, A resolution of the paradox of omniscience, in *Reason and Decision,* Bowling Green Studies in Applied Philosophy, vol. III-1981, ed. Michael Bradie and Kenneth Sayre (Bowling Green, OH: Applied Philosophy Program, Bowling Green State University, 1982), pp. 17–30.

7. By "strategy" I mean a course of action that can lead to any of the outcomes associated with it, depending on the strategy choice of the other player; the strategy choices of both players define an outcome at the intersection of their two strategies. While the subsequent moves and countermoves of players could also be incorporated into the definition of a strategy—meaning a complete plan of responses by a player to whatever choices the other player makes in the sequential game—this would make the normal (matrix) form of the game unduly complicated and difficult to analyze. Hence, I use "strategy" to mean the choices of players that lead to an initial outcome, and "moves" and "countermoves" to refer to their subsequent sequential choices, as allowed by rules II–IV.

is independent of the interpretation of the players as SB and P.

To illustrate this theory, assume each of the players chooses his compromise C strategy initially in Chicken, resulting in (3,3). If R (i.e., SB in Fig. 4.1) departed from this initial outcome and moved the process to (4,2), C (i.e., P in Fig. 4.1) could then move it to (1,1), and R could in turn respond by moving it to (2,4). These possible moves, and the corresponding "stay" choices at each node, are illustrated in Fig. 4.2.

To determine rational choices for the players to make at each node of the game tree, starting at (3,3), it is necessary to work backward up the game tree in Fig. 4.2. Consider R's choice at (1,1). Since R prefers (2,4) to (1,1), I indicate "stay" at (1,1) would *not* be chosen by a slash on its branch, should the process reach the node at (1,1).

Figure 4.2   *Game Tree for Moves, Starting with R, from (3,3) in Chicken*

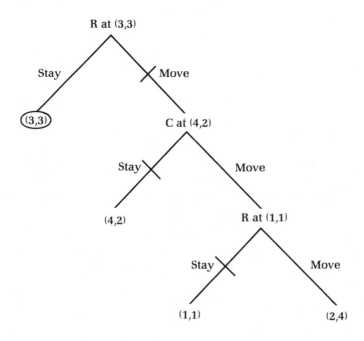

Key: (x,y) = (R,C)

4 = best; 3 = next best; 2 = next worst; 1 = worst

Circled outcome is rational—the outcome that "survives" the slashes

Instead, outcome (2,4) would be chosen, which can be mentally substituted for the endpoint, "R at (1,1)."

Working backward again, compare (4,2) at the "stay" branch with (2,4) at the "move" branch (given the previous substitution). Since C would prefer (2,4), the "stay" branch at this node is slashed, and (2,4) moves up to a final comparison with (3,3) at the top node. At this node R would prefer (3,3), so the "move" branch at the top node is slashed, and (3,3) is therefore the final outcome that "survives" the slashes.

In other words, there would be no incentive for R to depart from (3,3), anticipating the rational choices of players at subsequent nodes in the game tree of Fig. 4.2. Similarly, because of the symmetry of the game, there would be no incentive for C to depart from (3,3) in Chicken. When the final outcome coincides with the initial outcome, as it does in the case of (3,3), it is called a "nonmyopic equilibrium."[8]

For the other three outcomes in Chicken, there is no corresponding incentive for both players to stay at them, should any be the initial outcome of the game. For example, a game-tree analysis, starting at (1,1), reveals that R would have an incentive to depart to (2,4) and C to (4,2). After either departure, the process would terminate because the player with the next move—C at (2,4), R at (4,2)—would obtain his best outcome. But if (2,4) or (4,2) were the initial outcomes, rational departures by R from (2,4) and C from (4,2) would carry the process to (1,1), whence it would go to (4,2) if R departed initially, (2,4) if C departed initially, and stay for the reason just given.

But there is a complication starting at (1,1). Clearly, R would prefer that C depart first from (1,1), yielding (4,2), and C would prefer that R depart first, yielding (2,4). Since one cannot say *a priori* which player would be able to hold out longer at (1,1), forcing the other to move first, I indicate the final outcome, starting at (1,1), as "(2,4)/(4,2)"—either is possible. (Of course, if SB were one of the

8. Steven J. Brams and Donald Wittman, Nonmyopic equilibria in 2 × 2 games, *Conflict Management Peace Sci.* 6, 1 (1983); see also Marc D. Kilgour, Equilibria for far-sighted players, *Theory and Decision* (forthcoming), for an extension of the concept of nonmyopic equilibrium. To ensure that a final outcome is reached, either at the start or before there is cycling back to the initial outcome, the definition of a nonmyopic equilibrium also includes a termination condition. This condition specifies that there exists a node in the game tree such that the player with the next move can ensure his best outcome by staying at it. This condition is satisfied by the "cooperative" (3,3) outcome in both Chicken and Prisoners' Dilemma (discussed in Section 4.5).

players, he could presumably hold out longer than P, but in the present version of Chicken, I make no assumptions about one player's being more powerful—i.e., the Almighty in the Bible.)

It is easy to show that if (2,4) is the initial outcome, the final outcome according to the game-tree analysis would be (4,2), and (2,4) if (4,2) were the initial outcome. This is because the player obtaining his next-worst outcome (2), by moving the process to outcome (1,1), can force the other player to move to the outcome best for himself [(4,2) for R, (2,4) for C]. In either case, the player obtaining his best outcome (4) at (2,4) or (4,2) would seem to have no incentive to depart to the inferior outcome, (3,3).

Yet, an objection can be raised to this reasoning: the player who obtains 4 initially, knowing he will be reduced to 2, would have an incentive to move the process to (3,3) first, whereby he obtains his next-best outcome rather than his next-worst. Moreover, once at (3,3), the process would stop there since a subsequent move by, say, R to (4,2) would then move the process to (1,1), and thence to (2,4), which is inferior for R to (3,3).

This countermove to (3,3) by the player obtaining his best outcome at (2,4) or (4,2) would appear to introduce a new kind of rational calculation into the analysis—what the other player will do if one does not seize the initiative. True, I previously assumed that each player, separately, would ascertain the final outcome only for himself; yet, it will be recalled that in the earlier discussion of possible moves from initial outcome (1,1), I also assumed that each player would consider not only the consequences of departing from this outcome himself but also the consequences of the other player's departing. Because each player could do better by holding out at (1,1), I argued that each would strive to delay his departure, hoping to force the other player to move first.

The situation, starting at (2,4) or (4,2), is the reverse for the players. Although the game-tree analysis shows that, say, R should not move from (4,2) to (3,3), his recognition of the fact that C can move the process to (2,4) would induce him to try to get the jump on C by moving first to (3,3). In contrast, at (1,1) each player has an incentive to hold out rather than scramble to leave the initial outcome first.

In either event, a rational choice is dictated not only by one's own game-tree analysis but by that of the other player as well, which may cause one to override one's own (one-sided) rational choice. Henceforth, I assume that a final outcome reflects the *two-sided* analysis that both players would make of each other's rational choices, in addition to their own.

In the case of outcomes (2,4) and (4,2), it is impossible to say *a priori* which player would be successful in departing first. Accordingly, as in the case of (1,1), it seems best to indicate a *joint* final outcome of "(4,2)/(3,3)" starting from (2,4), and of "(2,4)/(3,3)" starting from (4,2).

In summary, the final outcomes of Chicken, given that the players make rational choices—according to a two-sided game-tree analysis and the four rules specified previously—are as follows for each initial outcome:

Initial Outcome	Final Outcome
(3,3)	(3,3)
(1,1)	(2,4)/(4,2)
(4,2)	(2,4)/(3,3)
(2,4)	(4,2)/(3,3)

If one substitutes the final outcomes for the initial outcomes in the payoff matrix of Fig. 4.1, the new game shown in Fig. 4.3 results. The outcomes of this game may be thought of as those that would be obtained if the four rules of sequential play specified earlier, coupled with rational choices based on a two-sided game-tree analysis, were operative.

In the preliminary analysis of this game, assume that each of the two outcomes in the joint pairs is equiprobable.[9] Then, in an expected-value sense, C dominates $\overline{C}$ for each player: if Column, for example, chooses C, (3,3) is better for Row than (2,4)/(3,3), which half the time will yield (2,4); if Column chooses $\overline{C}$, (4,2)/(3,3)

---

9. The equiprobability assumption is not crucial; it is made to illustrate the calculation of expected values and is contrasted with other assumptions given in the next paragraph.

Figure 4.3 *Revised Chicken, with Final Outcomes*

		Column	
		C	$\overline{C}$
Row	C	(3,3)	(4,2)/(3,3)
	$\overline{C}$	(2,4)/(3,3)	(2,4)/(4,2)

Key: $(x,y) = (R,C)$
  4 = best; 3 = next best; 2 = next worst; 1 = worst
  Circled outcome is a Nash equilibrium

is better for Row than (2,4)/(4,2) because, though the (4,2)'s "cancel each other out," (3,3) is preferred to (2,4) half the time.

Strictly speaking, however, for C to dominate $\overline{C}$ in *every play* of the game, it is necessary to make two assumptions: (1) whenever Column chooses $\overline{C}$, if $\overline{C}$ for Row yields (4,2) as a final outcome, so does C for Row; (2) there is some possibility, however small, that the choice of $\overline{C}\overline{C}$ by the players yields (2,4). In this manner, Row's choice of C is always at least as good as, and sometimes better than, choosing $\overline{C}$.

Assumption (1) above is the crucial assumption. It says, in effect, that whenever Column chooses $\overline{C}$ in the original game of Chicken, and Row can hold out longer at (1,1) if he chooses $\overline{C}$ himself—forcing the final outcome to be (4,2)—Row can preempt Column at (2,4) if he chooses C, yielding the final outcome (4,2). In other words, if R is the "stronger" player at (1,1), he is also the "quicker" player at (2,4), because he is able to move the process to (1,1) before Column moves it to (3,3).

The guarantee of dominance provided by assumptions (1) and (2) seems as reasonable as the expected-value assumption which says, given the equiprobability of the two outcomes in the joint pairs, that C dominates $\overline{C}$ "on the average." Either way, rational players in Chicken, anticipating the final outcomes shown in Fig. 4.3, will each choose their dominant strategy C. Thereby the four rules of sequential play specified earlier induce the cooperative (3,3) outcome in Chicken, which is circled in Fig. 4.3.

In this way, the paradox of omniscience is resolved in Chicken because, given the rules of sequential play, SB's omniscience confers on him no special advantage. In fact, the (3,3) resolution is exactly the same under the specified rules if neither player is omniscient, thus rendering the result stable, independent of either player's omniscience.

There is no such "fair" resolution—that is, a dominant-strategy outcome ranked the same by both players—for the five other 2 × 2 ordinal games vulnerable to the paradox of omniscience. The outcome matrices of these games are shown in Fig. 4.4, with both their initial and final outcomes.[10] Note that if the row player

10. Of the 78 structurally distinct 2 × 2 ordinal games—in which each player can strictly rank the four outcomes from best to worst, and no interchange of rows, columns, or players can transform one of these games into any other—these five games plus Chicken are the only games in which neither player has a dominant strategy, associated with the outcome induced by SB's omniscience, that is best for P (column player in Fig. 4.4) and inferior for SB (row player). Accordingly, SB's omniscience, and P's awareness of it, are required to guarantee its

Figure 4.4   *Five Games, Other Than Chicken, Vulnerable to the Paradox of Omniscience*

Initial Outcomes		Final Outcomes	
(4,3)	(2,1)	(3,4)	(3,4)/(4,3)
(1,2)	(3,4)	(4,3)/(3,4)	(4,3)
(4,2)	(2,1)	(3,4)	(2,4)/(4,3)
(1,3)	(3,4)	(4,3)/(2,4)	(4,2)
(4,3)	(1,1)	(2,4)	(3,4)/(4,2)
(3,2)	(2,4)	(4,2)/(4,3)	(4,3)
(4,3)	(1,1)	(3,4)	(4,3)/(3,4)
(2,2)	(3,4)	(3,4)/(4,3)	(4,3)
(4,3)	(2,2)	(3,4)	(3,4)/(4,3)
(1,1)	(3,4)	(4,3)/(3,4)	(4,3)

Key: 4 = best: 3 = next best; 2 = next worst; 1 = worst
Circled outcomes are Nash equilibria

is an omniscient SB and the column player P, in each of the initial-outcome payoff matrices, P, by choosing his second strategy, forces SB to choose a strategy that results in the best outcome for P but an inferior outcome for SB.

In the case of the final-outcome payoff matrices, however,

choice. However, there are two other 2 × 2 games in which P can force SB to choose an outcome of lower rank than his; but this outcome, (3,2), is the next-best, not best, for P. Furthermore, in each of these games, there is another outcome, (3,4), better for *both* players but, paradoxically, unattainable if SB is omniscient and P knows this. For complete listings of the 78 2 × 2 games, see Anatol Rapoport and Melvin Guyer, A taxonomy of 2 × 2 games, *General Systems: Yearbook of the Society for General Systems Research* 11 (1966), 203–214; and Brams, Deception in 2 × 2 games. The six games vulnerable to the paradox of omniscience, and the two in which P can beat SB but not obtain his best outcome, are identified by their numbers in each listing in Brams, Mathematics and theology, p. 281.

the circled outcomes are the product of dominant strategy choices by both players, whether omniscient or ordinary. Unlike Chicken, these dominant strategies depend only on the assumption that there is some possibility, however small, that *each* of the outcomes in the joint pairs will occur. It is not necessary to make either an equiprobability assumption or an assumption about the conditional occurrence of events, as in Chicken, to ensure dominance in these other paradoxical games.

Yet, there is a price to pay in indeterminacy for the "resolution" achieved in these games. First, although each player has a rational strategy choice—his dominant strategy associated with the circled outcomes—these outcomes are all joint pairs.[11] Moreover, one of the two outcomes favors one player, the other outcome the other player, so the resolution is necessarily "unfair" to one player. In general, the disadvantaged player is the one who "gives in" first by moving from a Pareto-inferior outcome in the original (initial-outcome) game.

Perhaps, though, there is a kind of cold-blooded justice in these resolutions. Observe that they always ensure at least the security level of 2 for each player, which is the best outcome a player can ensure *by himself for himself* (by not choosing his strat-

---

11. The (single) final outcomes along the main diagonal reflect the fact that the player not receiving his best outcome can move the process "through" an off-diagonal inferior outcome—for both players—to the other diagonal outcome best for himself. However, one might take a different view of what constitute final outcomes along the diagonal, which the previous game-tree analysis does not show because it terminates before outcomes are repeated.

More specifically, the player initially receiving his best outcome can always trigger a series of moves from this outcome that will bring the process back to it. Moreover, he would be motivated to do so to prevent the other player from moving the process to an outcome not best for him. Thus, a two-sided analysis of these five games suggests that *both* players would have an incentive to move first from the diagonal outcomes (if these are the initial outcomes), thereby transforming them into joint pairs as final outcomes.

However, this view assumes that a player would be motivated to move from an initial outcome simply to ensure that it returns to this starting point—but now as a final outcome. Yet this kind of calculation, though consistent with the earlier rules, seems a bit artificial. In any event, whether the final outcomes along the diagonal are considered single outcomes (as in Fig. 4.4) or joint pairs, it would be a joint pair that would be the final outcome of the game under either interpretation. Hence, the alternative view on final outcomes along the diagonal—that they are joint pairs—would have no effect on the resolution of the paradox: because *all* outcomes would be the same in the final-outcome matrix under the alternative interpretation, *any* choice of strategies by the players would yield the dominant-strategy final outcomes circled in Fig. 4.4.

egy associated with his worst outcome of 1). Also, in four of the six paradoxical games (including Chicken), the resolutions give both players at least their next-best outcomes of 3. That is the justice in these resolutions. The injustice in the five paradoxical games other than Chicken—if there is any—is that the stronger or more powerful player, able to force the other player to move first from an inferior outcome in the original game, always benefits, relatively speaking, by obtaining his best outcome of (4). I shall return to this and other notions of power in Chapters 5 and 6.

Structurally, Chicken is the only *fair game* among the six, conferring no special advantage on the stronger player (if there is one). At the same time, however, Chicken requires more stringent assumptions to ensure dominance in the final-outcome matrix and, therefore, the (3,3) resolution. Since one set of dominance assumptions for Chicken discussed earlier has a kind of strength-and-quickness-of-player interpretation, even Chicken seems not to be immune to a resolution that, while fair, indicates the presence of a stronger and quicker player.

In the other five paradoxical games, the stronger player has an advantage in determining which one of the two outcomes, in the circled joint pairs, is actually selected. The fact that the other player is not severely hurt, obtaining at least his security level, suggests that the rules of the game are not unjust.

In general, except for Chicken, these rules resolve the paradoxical games in favor of the stronger player without, at the same time, woefully depriving the weaker player. This is better than providing no resolution—and, potentially, disaster in games like Chicken—as a consequence of the absence of dominant strategies in the initial-outcome games. Allowing rational sequential moves after initial outcomes have been chosen seems not only to mirror the reality of ongoing games but also, at least in the paradoxical games, obviates disastrous outcomes for either one or both players.

One curiosity, perhaps, about the resolution of the five games vulnerable to the paradox of omniscience in Fig. 4.4 is that they all entail the rational choice of a Pareto-inferior *initial outcome*. Only after this choice does a move by Row or Column from this outcome lead to a Pareto-superior final outcome that resolves the paradox. In Chicken, on the other hand, the resolution leads to the initial choice of the cooperative (3,3) outcome, from which neither player has an incentive to move because it is a nonmyopic equilibrium (see note 8).

4.5. A Second Paradox: Moves May Provide No Resolution

The sequential rules of play postulated in Section 4.4 not only resolve the paradox of omniscience in Chicken and the five other paradoxical games but also induce a Pareto-superior outcome that is not the worst for either player. This outcome, except in Chicken, benefits the stronger of the players, based on their comparative rankings of the outcomes; in later chapters, I shall formalize this superior ability in terms of SB's omnipotence and related attributes he might possess.

A player's omniscience, however, and his opponent's knowledge that he possesses this predictive power, may lead to a second paradox that is not so easily resolved. Consider the Testing Game in Fig. 4.5, in which SB must choose either to test or not test P, and P may either prepare or not prepare for a possible test.[12]

This kind of game is ubiquitous in the Bible. For example,

---

12. This is one of six games vulnerable to the second paradox of omniscience (to be discussed below); it and the other five games are identified in Steven J. Brams, Omniscience and omnipotence: how they may help—or hurt—in a game, *Inquiry* 25, 2 (June 1982), 217–231.

Figure 4.5 *Outcome Matrix of Testing Game*

		P	
		Prepare	*Don't prepare*
Test		P accurately anticipates test (1,4)	←	P inaccurately anticipates no test (4,1)
Don't test		P inaccurately anticipates test (2,2)	→	P accurately anticipates no test (3,3)

SB (left side)

Test — P accurately anticipates test (1,4) ← P inaccurately anticipates no test (4,1)

↓   ↑

Don't test — P inaccurately anticipates test (2,2) → P accurately anticipates no test (3,3)

↑

Optimal strategy of P if he is aware that SB is omniscient

Key: $(x,y) = (SB,P)$

4 = best; 3 = next best; 2 = next worst; 1 = worst

Arrows between outcomes indicate rational departures from each outcome by SB (vertical) and P (horizontal)

Abraham's faith is tested by God when God commands him to sacrifice his son, Issac. As another example, Job is tested by Satan, with God's approval, when he is subjected to a series of calamaties that only stop short of his annihilation.

In the Testing Game I assume that SB does not want P apprised of a possible test; otherwise, its results could be invalidated. Thus, in Genesis 22:1 we are told that "God put Abraham to the test," but Abraham is never informed that the sacrifice he thinks he is about to perform is only a test. (That Abraham passed the test "unprepared" very much pleased God, though presumably Abraham would have preferred not to be confronted with such a harrowing situation, even though he was rewarded in the end for remaining faithful.) In the dispute between the two women over the maternity of a baby that Solomon had to adjudicate (I Kings 3:16–28), if the impostor had known that Solomon's threat to cut the baby in half was designed to ascertain which woman would give up her claim and thereby identify herself as the mother, she (the impostor) might have protested Solomon's proposed solution, too, which would have ruined the test.

I assume the following goals of the players in the Testing Game:

SB: (1) Primary goal—prefers P not prepare over prepare (making a valid test possible);

(2) Secondary goal—prefers to test P when he is not prepared, not to test him when he is;

P: (1) Primary goal—prefers accurately to anticipate test/no test

(2) Secondary goal—if he anticipates a test, prefers test; if not, prefers no test.

This is clearly a game of partial conflict, with one Pareto-inferior outcome, (2,2), one "cooperative" outcome, (3,3), and two outcomes, (4,1) and (1,4), in which the preferences of the two players are diametrically opposed.

Now assume that SB is omniscient and P knows this. Then P, if rational, will choose preparation, which induces SB not to test and yields (2,2). Otherwise—had P chosen not to prepare—SB would have predicted this choice and exploited this strategic information to obtain (4,1), his best outcome and P's worst.

The paradox here is not that SB is hurt, relative to P, by his omniscience and P's knowledge of it; rather, *both* players are hurt by the resulting Pareto-inferior (2,2) outcome. They would both prefer the Pareto-superior (3,3) outcome, as in Prisoners' Dilemma, but SB's omniscience, and P's knowledge of it, undermine this "cooperative" outcome.

In the absence of omniscience, it is difficult to say what outcome would be chosen by the players in the Testing Game. Neither has a dominant strategy, and, unlike Chicken, there are no Nash equilibria. For example, if (3,3) were chosen initially, SB would have an incentive to defect to his first strategy of testing, as indicated by the vertical arrow from (3,3) to (4,1).

But the movement does not end here. P has an incentive to move the process from (4,1) to (1,4), as indicated by the horizontal arrow between these outcomes. To complete the cycle, SB would move the process from (1,4) to (2,2), and P would return it to (3,3). In a word, there is no stability in this game.

Contrast this instability with the evident stability in the Prisoners' Dilemma of the fourth Knowability Game (Fig. 2.3), reproduced in Fig. 4.6, in which the arrows converge to (2,2) in the initial-outcome matrix. The players' dominant strategies associated with this outcome, which is a Nash equilibrium, would suggest that this is the unique stable outcome in this game.

But recall the four sequential rules of play that I applied to Chicken and the other paradoxical games in Section 4.4. The game tree for Prisoners' Dilemma, shown below the initial-outcome and final-outcome matrices in Fig. 4.6, demonstrates that, thinking ahead, neither player would be motivated to depart from (3,3), because if he did [say, R moved to (4,1)], the other player (C) would countermove to (2,2), where subsequent moves would terminate.

This is so because if R moved from (2,2) to (1,4), C would stay since by doing so he could implement his best outcome. But since R can anticipate that the process would end up at (2,2) if he departed initially from (3,3), he would have no incentive to depart from (3,3) in the first place. Thereby (3,3) is the rational outcome when Prisoners' Dilemma is played according to the sequential rules.

The two-sided analysis I illustrated in the case of Chicken in Section 4.4, when applied to Prisoners' Dilemma, shows that from initial outcomes (4,1) and (1,4), the process would move to final outcome (3,3). At (4,1), for example, it would be in *both* players' interest that R move to (3,3) before C moves to (2,2), where the process would stop, as I indicated in the previous paragraph. Because there is no incentive for R to move to (1,4) or C to move to (4,1) from (2,2), it, like (3,3), is stable in a nonmyopic sense.

Thus, were (2,2) initially chosen by the players, it would be the final outcome, whereas all other outcomes would be trans-

Figure 4.6 *Fourth Knowability Game (Prisoners' Dilemma)*

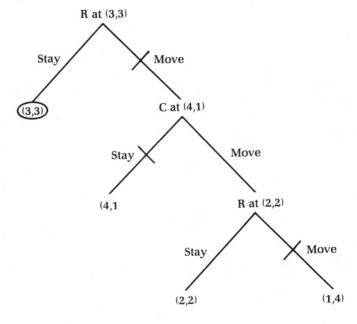

| **Initial Outcomes** | | | | **Final Outcomes** | |
| C | | | | C | |

	(1,4)	←	(3,3)		(3,3)	(3,3)
R	↓		↓	R		
	(2,2)	←	(4,1)		(2,2)	(3,3)

R at (3,3)

Stay / Move

(3,3)

C at (4,1)

Stay / Move

(4,1)

R at (2,2)

Stay / Move

(2,2)   (1,4)

Key: $(x,y) = (R,C)$

4 = best; 3 = next best; 2 = next worst; 1 = worst

Arrows between outcomes in initial-outcome matrix indicate rational departures from each outcome by R (vertical) and C (horizontal)

Circled matrix outcomes are dominant-strategy Nash equilibria; circled outcome in game tree is "surviving" outcome

formed into (3,3). Altogether, the final-outcome matrix of Prisoners' Dilemma given in Fig. 4.6 shows only the upper-right (3,3) Nash equilibrium to be the product of dominant strategy choices by the players and presumably the outcome that would be chosen. Note that it coincides with the (3,3) outcome in the initial-outcome matrix; the other two (3,3) outcomes in the final-outcome matrix are Nash equilibria but not ones associated with the players' dominant strategies.

Thus, like Chicken, the sequential rules lead to the cooperative (3,3) outcome in Prisoners' Dilemma, but, unlike Chicken, the implementation of this outcome does not depend on one player's being stronger and quicker than the other. Not only does the dominance of strategies associated with (3,3) in the final-outcome matrix require no special assumptions, but the dominance of the players' other strategies, associated with (2,2) in the initial-outcome matrix, is reversed.

The cooperative solutions that the sequential rules provide in Chicken and Prisoners' Dilemma, however, do not extend to the Testing Game. Not only does the cycling I described earlier have no similar resolution at (3,3), but there is also no termination at any other outcome. Hence, one cannot even construct a final-outcome matrix for this game. Only a weaker notion of stability, based on the concept of an "absorbing outcome," suggests that (3,3) might be chosen by rational players in this game.[13]

To be sure, as I shall show in Chapter 5, omnipotence enables SB to implement (3,3) and thereby overcome the difficulty his omniscience precipitates in the Testing Game. Barring omnipotence, however, the anomaly of a Pareto-inferior (2,2) outcome in this game that SB's omniscience, and P's knowledge of his omniscience, induces seems to me to be something not easily brushed aside.

In particular, the second paradox of omniscience suggests P should remain prepared, expecting to be tested but, strangely, know that testing is not rational for SB. The theological consequences of these choices would seem to be that one should worship God and seek His forgiveness—that is, be prepared to be tested, asking for His mercy in advance. Yet, if one believes Him to be omniscient, these very signs of faith will make His testing unnecessary (presuming P is seen as credible—not faking his piety), whereas the unrepentant can expect to be tested, with presumably unsatisfactory results because they are unprepared.

That God benefits most (4) from testing sinners who, being unprepared, fail His test may be disputed. So may God's relatively poor outcome when He chooses not to test the faithful/prepared. But, in my opinion, there is no disputing that God's presumed

---

13. Steven J. Brams and Marek P. Hessel, Absorbing outcomes in 2 × 2 games, *Behavioral Sci.* 27, 4 (October 1982), 393–401.

omniscience may provoke considerable anxiety and concern among many believers.

In fact, I conjecture that it is no accident that atonement and confession are most prominent in religions that stress the constant need for vigilance against sin and temptation. In these religions, an unending series of tests of faith, which are difficult if not impossible to pass, dictate that one's efforts and preparation be redoubled to avoid the most dire consequences.

Put another way, as unsatisfactory as (2,2) is for P (as well as SB) in the Testing Game, it is still better for P than (4,1), given his recognition of SB's omniscience. One question I shall analyze in Chapter 5 is how SB's omnipotence can release the players from the (2,2) trap that SB's omniscnience induces in this game.

4.6. Conclusions

The first paradox of omniscience demonstrated that the possession of omniscience by SB, and P's awareness of it, can, in games like Chicken (Fig. 4.1), help P and hurt SB. The four sequential rules of play, on which the theory of moves is based, may rescue SB from the unfortunate outcome in such games and may even swing the balance in his favor if he is the stronger and quicker player. This is not to say he is omnipotent, in the sense of being all-powerful, but rather that he has certain superior abilities, to be described more systematically in Chapters 5 and 6.

The new sequential rules suggested a way for players to look ahead in a game and calculate optimal moves from each outcome, based on a backward-induction analysis of the game tree of possible moves. This enables one to determine a final-outcome matrix when there are stable outcomes to which the move-countermove process gravitates. Nash equilibria in this matrix generally coincide with nonmyopic equilibria; in the case of Chicken and Prisoners' Dilemma (Fig. 4.6), the cooperative (3,3) outcome is a nonmyopic equilibrium, but it is not a Nash equilibrium in the initial-outcome matrix of either game.

In the Testing Game (Fig. 4.5), whose initial-outcome matrix contains no Nash equilibrium, the players cycle over the four outcomes. This prevents a theory-of-moves resolution of the second paradox of omniscience that induces (2,2) in this game, which is inferior for both players to the cooperative (3,3) outcome.

The (2,2) outcome suggests how one's perception of God's omniscience may lead one to be devout in order to guard against an unanticipated test of faith, and why God would then not feel it necessary to put one to the test. The preferable (3,3) outcome for both players would leave P unprepared in the Testing Game, vulnerable to a strategy switch by SB. How SB might successfully implement this and other unstable outcomes with his omnipotence will be explored in Chapter 5.

FIVE

Omnipotence: Moving and Staying Power

5.1. Introduction

The resolution of the first paradox of omniscience in Chapter 4, via the theory of moves, was thoroughly secular. It generally favored the stronger or quicker player, in a particular sense, and thus could be viewed as unfair to the other player, though he always obtained at least his security level in a game.

But an asymmetry between players is a fundamental tenet of most religions, which have populated the world with a host of figures at varying levels in their hierarchical orders, including God, Christ and other messiahs, Satan, angels, prophets, judges, saints, pagans, infidels, and numerous other believers and nonbelievers. Each of these figures is endowed with particular attributes, but probably the most striking differences between them are their different powers.

In Chapter 4, strength and quickness were introduced "through the back door," so to speak, to illustrate conditions under which one player would have a dominant strategy in certain games. In this chapter, I shall define two different concepts of power—moving and staying—and delineate some of their consequences in games played between SB and P. A third concept, that of threat power, will be reserved for Chapter 6 because it seems to be most closely associated with SB's immortality.

The advantage that the concepts of moving and staying power offer is that they can be precisely defined in elementary games of the kind analyzed here. However, they are probably not what most philosophers and theologians think of as "omnipotence," which literally means power over all.

The Bible, in my opinion, makes clear that God's powers are limited, just as I suggested His omniscience may be only partial.[1] First, except in a very few instances, such as when God "harden[s] Pharaoh's heart" to make him obstinate in the face of demands that the Israelites be freed from Egyptian bondage (Exod. 7:3), biblical characters seem generally able to exercise their free will and retain their autonomy. Second, in the narratives themselves, God often does not get His way and, in fact, on more than one occasion threatens to destroy the world He created because His goals are frustrated or His hegemony is challenged.

But, of course, God is not impotent. He demonstrates His prowess by performing miracles, punishing and sometimes exterminating the wicked, and helping the righteous. While esoteric questions and metaphysical riddles that philosophers have tied to His omnipotence may not have ready answers,[2] neither are such questions and riddles, in my view, central to understanding the significant implications entailed by the power of a SB. Rather, I

---

1. Martin Gardner rejects the view that there are limitations: "Let me confess at once that I find something profoundly impious, almost blasphemous, about setting limits of any sort on the power of God to bring things about in any manner that He chooses." Martin Gardner, *The Ambidextrous Universe: Mirror Asymmetry and Time-Reversed Worlds*, 2nd ed. rev. (New York: Scribner's, 1979), p. 125.

2. For recent investigations into such questions in the philosophy of religion, see *The Power of God: Readings on Omnipotence and Evil*, ed. Linwood Urban and Douglas N. Walton (New York: Oxford University Press, 1978); Anthony Kenny, *The God of the Philosophers* (Oxford: Clarendon, 1979); Gary Rosenkrantz and Joshua Hoffman, What an omnipotent agent can do, *Int. J. Philos. Religion* 11, 1 (Spring 1980), 1–19; W. S. Anglin, Can God create a being He cannot control? *Analysis* 40, 4 (October 1980), 220–223; and Edward Wierenger, Omnipotence defined, *Philos. and Phenomenological Res.* 43 (1982) (forthcoming). None of this literature analyzes the *relationship* between a superior and ordinary being as a game, which in my view puts the question of omnipotence, as well as omniscience, in a fundamentally new framework. Omnipotence, by the way, seems to have confounded philosophers more than omniscience, perhaps because "the notion of infinite power has seemed too obscure, too shrouded in mystery and ineffability for us to analyze our feelings of awe and bring them into the domain of pure concepts." *The Power of God*, p. 4.

prefer to start with some intuitively reasonable definitions of power/omnipotence and try to ascertain in what kinds of games SB's possession of this power affects the outcomes he can realize.

5.2. Moving Power[3]

Consider again the Testing Game (Fig. 4.5), described in Section 4.5 and shown again in Fig. 5.1 (ignore for now the arrows, both single and double, between pairs of outcomes in this representation). Recall that the second paradox of omniscience produced the following result: in light of SB's predicted best response because of his omniscience, P should choose "prepare" [resulting in (2,2)] over "don't prepare" [resulting in (4,1)] to maximize his payoff.

In this game, as in the Revelation Game described in Section 2.2 (Fig. 2.1), and also reproduced in Fig. 5.1, the Pareto-superior (3,3) outcome [(3,4) in the Revelation Game] would be implemented if there were a role reversal—P was omniscient and SB was not but anticipated his (P's) best response to his strategy choice. In these games, then, if SB had the power somehow to transfer his omniscience to P, it would be in his interest to do so, for he, as well as P, would do better at (3,3) or (3,4).

---

3. Material in this and the next section is based largely on Steven J. Brams, Omniscience and omnipotence: how they may help—or hurt—in a game, *Inquiry* 25, 2 (June 1982), 217–231.

Figure 5.1 *Outcome Matrices of Revelation and Testing Games*

		Revelation Game P				Testing Game P	
		Believe	**Don't believe**			**Prepare**	**Don't prepare**
SB	**Reveal**	$(3,4)^P$ ←	(1,1)		**Test**	(1,4) ←	(4,1)
		⇓	⇑	**SB**		⇓	⇑
	Don't reveal	$(4,2)^{SB}$ →	(2,3)		**Don't test**	(2,2) →	$(3,3)^{SB,P}$

Key: $(x,y) = $ (SB,P)

4 = best; 3 = next best; 2 = next worst; 1 = worst

Double arrows between outcomes indicate moves of player with M-power (assumed to be SB in this figure); single arrows, moves of player (P) without M-power

Superscripted outcomes indicate those that player with M-power (SB or P) can implement

These games illustrate the benefits of omniscience (strangely, for P in these cases), whereas the paradoxes of omniscience in Chapter 4 illustrated its costs (to SB in Chicken and other games, to both players in the Testing Game). On balance, it is hard to say whether omniscience is helpful or harmful or neither [recall that, in the Revelation Game, SB's omniscience had no effect on the Pareto-inferior (2,3) Nash equilibrium]. The effects of moving power, as I shall show next, are not so variable: while they may not always be helpful, they are never harmful.

It is difficult to conceptualize the exercise of power in a game in which two players—whether one is omniscient or not—make simultaneous strategy choices, as usually assumed of games in normal (matrix) form. To introduce a notion of power into game playing, it is useful to resurrect the rules described in Section 4.4 that allow for a sequence of moves and countermoves by the players *after* the initial (simultaneous) strategy choices are made. Given these rules are operative, I assume

   Nonmyopic Calculation: Players make choices in full anticipation of how each will respond to the other, both in selecting their strategies initially and making subsequent moves.

Because rules I–IV are symmetric with respect to the players, and Nonmyopic Calculation, according to these rules, does not distinguish one player from the other, neither player can be said to be more powerful than the other on the basis of his move-countermove capability.

Now assume that one player is M-powerful (has moving power) and the other is not. Although I assume rules I–IV still hold for the two players, the ability of the other (M-limited) player to "choose" to switch his strategy (rule IV) is assumed to be constrained:

   A *M-powerful* player can continue to move indefinitely. If he
   does, the *M-limited* player must stop after some finite number of moves. This number need not be prespecified, so long
   as it exceeds three or four moves that would allow for the
   possibility of cycling (as shown below).

As before, under rule IV, when *either* player whose turn it is does not switch his strategy, the outcome reached is final.

Note that the M-power of one player does not mean that it is the M-limited player who *must* stop the process; the M-powerful player may cease moving, too. However, if he does not, then it becomes obligatory for the M-limited player eventually to desist from making a move. Clearly, this constraint on the M-limited player's ability to continue moving injects an element of asymmetry into where the "alternating responses," posited in rule IV, will terminate, mirroring the greater power of the M-powerful player.

Does M-power really capture the essence of "omnipotence"? In itself, almost certainly not, and that is why I shall introduce other kinds of power later. But it does reflect a kind of indefatigability: though the M-powerful player cannot force a particular choice on the M-limited player—consistent with allowing the players free will, or independent choices—he can hold out longer and thereby induce the M-limited player to concede.

The notion of M-power is implicit in several stories in the Bible. For example, in Moses's and God's confrontation with Pharaoh in the Book of Exodus, it is the former players who eventually, after visiting devastating plagues on Pharaoh, coerce him (after the tenth plague) into letting the Israelites go, albeit only temporarily.

To illustrate this notion of omnipotence, consider the Testing Game in Fig. 5.1. Assume SB is M-powerful and (3,3) is the initial outcome. It would seem that P would have no incentive to move from (3,3) to (2,2) because he would do worse if he did; moreover, the process would stop there, for if SB subsequently moved to (1,4), P would certainly stay there since this is his best outcome while it is SB's worst.

Now consider SB. As the M-powerful player, would he move from (3,3) to (4,1)? If he did, P, obtaining his worst outcome, would move to (1,4). But then SB would move to (2,2), and P would return the process to (3,3). The cycle would be complete, as indicated by the arrows.

The double arrows (vertical) indicate that SB can *always* move from (3,3) and from (1,4). P, recognizing this, could stop the process either at (2,2) or (4,1), the endpoints of these arrows where it is his next move. Obviously, it would be in his interest to choose (2,2), though he, as well as SB, would prefer (3,3). Since both players know this, I assume that the process will terminate at (3,3), even though, theoretically, SB has the power to continue to move from this outcome.

He would not, though, because the better outcome at which P could terminate the process, (2,2), is worse for him as well as P; hence, it is in SB's interest to effect termination of the move-countermove process at (3,3). Thereby, SB's M-power would induce (3,3) as the *final outcome*, which I assume in the theory of moves is the only outcome of consequence to the players (i.e., intermediate outcomes contribute negligible value to the players).

It is a bit odd, perhaps, to argue that SB's M-power is crucial in the Testing Game since it is SB, despite his *moving* power, who *stopped* play. But he does so to prevent the implementation of (2,2) by P, who would have had to stop eventually if SB did not.

The case for the effectiveness of M-power is more persuasive in the Revelation Game in Fig. 5.1. In this game, P can stop the process either at (4,2) or at (1,1), which are, as before, the endpoints of the double arrows in this game. Of course, P would prefer (4,2), but this is still only his next-worst outcome. Can he do better for himself [at, e.g., (3,4), which is Pareto-superior to the (2,3) Nash equilibrium]?

The answer is no, because SB can (and will) always move from (3,4) to (4,2) since the latter outcome is his best. Unlike the Testing Game, SB, by virtue of his M-power, does not have to settle for next best.

I have assumed that SB is the M-powerful player in both the Testing and Revelation Games. It is not very plausible, I suppose, to argue that P might have M-power, but if he did and SB were the M-limited player, the consequences would vary.

In the Testing Game, P would be able to do no better than SB, inducing (3,3), based on their comparative rankings. P's M-power would permit SB to stop the process either at (1,4) or (3,3), and SB would prefer the latter. On the other hand, in the Revelation Game, P could force on SB a choice between (3,4) and (2,3); clearly, it would be in both players' interest that (3,4) be implemented, which is P's best outcome. In summary, each player, if M-powerful, can obtain his best outcome in the Revelation Game but only his next-best in the Testing Game.

This conclusion does not depend on the initial outcome of either the Revelation or Testing Game. After the players have made their simultaneous strategy choices, according to rule I, which defines the initial outcome, at least one player has an incentive to move from that outcome, whatever it is, as allowed by rule II and illustrated by the arrows in Fig. 5.1. This is most apparent in the Testing Game, in which the player moving from an outcome, as indicated by the arrow emanating from that outcome, stands to gain immediately by switching to his other strategy.

In the case of the Revelation Game, however, the benefit from moving is not always immediate. Consider outcome (2,3), from which the departing arrow says that SB would move to (1,1). Of course, the reason for this move is not that he gains at (1,1) but rather that he forces P, who also suffers his worst outcome at (1,1), to move to (3,4). In other words, it is rational for SB to move from

(2,3) because he anticipates a (rational) countermove, according to rule III, by P to (3,4)—better for both players than (2,3)—though (3,4) itself is not stable.

The rationality of such moves can be determined from the game-tree analysis described in Chapter 4 for Chicken (Fig. 4.2) and Prisoners' Dilemma (Fig. 4.6). These games, with single and double arrows starting at initial outcome (3,3), are shown in Fig. 5.2. [The representation of Prisoners' Dilemma is different from that in Figs. 2.3 and 4.6: not only have the strategies been relabeled, but the two column strategies have also been interchanged to make it appear more comparable to Chicken in Fig. 5.2—with (3,3) in the upper left-hand corner—but structurally the game is the same.]

In the case of Chicken, SB's departure from (3,3) would carry the process to (2,4), where it would terminate because, at this point, P obtains his best outcome and, therefore, would have no incentive to move further. In the case of Prisoners' Dilemma, the

Figure 5.2 *Outcome Matrices of Chicken and Prisoners' Dilemma*

Chicken

		P	
		Compromise	*Don't compromise*
SB	*Compromise*	(3,3)*	((2,4))
		⇓	⇑
	Don't compromise	((4,2)) →	(1,1)

Prisoners' Dilemma

		P	
		Compromise	*Don't compromise*
SB	*Compromise*	(3,3)*	(1,4)
		⇓	
	Don't compromise	(4,1) →	((2,2))

Key: (x,y) = (SB,P)

4 = best; 3 = next best; 2 = next worst; 1 = worst

* = M-power outcome

Circled outcomes are Nash equilibria

Arrows indicate where process will go if SB (with M-power) moves from upper left–hand outcome process would get only as far as (2,2), for reasons already given in Section 4.5: in the Fig. 5.2 representation, if SB moved from (2,2) to (1,4), P would terminate the process at (1,4), which is worse for SB than (2,2).

In either case, SB's departure from (3,3) would trigger a countermove (and subsequent move in the case of Chicken), leading to a worse outcome (2) for SB in both games. That SB has M-power, then, does not upset the nonmyopic stability of (3,3) in either Chicken or Prisoners' Dilemma—or, for that matter, the nonmyopic equilibria in any of the 37 (of 78) distinct 2 × 2 ordinal games that have them (see Appendix for a partial listing).

Because neither (3,3) in the Testing Game, nor (3,4) in the Revelation Game, is a nonmyopic equilibrium, the M-power of one player *is* consequential in these games. In effect, M-power renders one outcome "final" in these games, according to rule IV, which it would not be in the absence of M-power because the players would continue to cycle indefinitely over the four outcomes. The M-power of one player precludes this from happening because the M-limited player will have to stop the process unless it is rational (as in the Testing Game in Fig. 5.1) for SB to stop the process first [at (3,3) in this game].

When it is not in his interest to do so, as in the Revelation Game—wherein SB can force (4,2)—the M-powerful player exercises "effective power." In fact, I define a player's M-power to be *effective* when he can ensure for himself a better outcome than when the other player has this power. This is the case in the Revelation Game, in which each player can induce his best outcome (4) if he is M-powerful; if the other player has this power, he is forced to accept an inferior outcome (2 for P, 3 for SB). In contrast, because each player's possession of M-power leads to the implementation of the same (3,3) outcome in the Testing Game, M-power is ineffective in this game.

It turns out that M-power is effective in 18 of the 78 distinct 2 × 2 ordinal games. In these games, it clearly matters which player is M-powerful, for he would suffer if the other player were. In the Revelation Game, for example, SB's M-power leads to an outcome worse for P [(4,2)] than if the Pareto-inferior Nash equilibrium [(2,3)] were chosen, not to mention the Pareto-superior (3,4) outcome that P's M-power would induce.

M-power *always* induces Pareto-superior outcomes in a 2 × 2 game, whether M-power is effective or ineffective. Thereby it provides an escape from the Pareto-inferior (2,2) outcome in the Testing Game that SB's omniscience induces.

There is, however, an important difference between games in which M-power is effective and ineffective. In the former games, the M-powerful player *always* does at least as well, and sometimes better, than the M-limited player, as measured by their comparative rankings of the outcomes. Thus, in the Revelation Game, each player obtains his best outcome when he is M-powerful, whereas the other player obtains an inferior outcome.

5.3. Is There a Paradox of Moving Power?

As noted in Section 4.4, nonmyopic equilibria are final outcomes, as defined by rules I–IV: from these outcomes, neither player would depart because, anticipating the consequences of moves and countermoves set off by his departure, he could expect to end up at an outcome inferior to the nonmyopic equilibrium. Thus, in games with nonmyopic equilibria, as illustrated in the case of Chicken and Prisoners' Dilemma in Section 5.2, M-power does not alter the fact that an M-powerful player's departure from a non-myopic equilibrium would eventually lead to a worse outcome for him, from which the M-limited (and perhaps the M-powerful player, too) would have no incentive to move [e.g., an irrational departure from (3,3) in Prisoners' Dilemma would eventually lead to (2,2), from which neither player would move]. Hence, M-power has no effect on the stability of nonmyopic equilibria.

Given that M-power confers no special advantage on players in games with nonmyopic equilibria, one might ask whether M-power can ever hurt a player. By "hurt" I mean that M-power can induce an outcome worse for the M-powerful player than the M-limited player (akin to the paradox of omniscience).

The answer is yes—but with qualifications. There are several games without nonmyopic equilibria in which, if SB is the M-powerful row player, the outcome induced is (3,4) or (2,3), and hence ranked higher by P. Two examples of such games are given in Fig. 5.3. In Game 1, if SB is M-powerful, he can always move the process from (3,4) to (2,3), but it would not be in his interest to do so. For at (2,3), if P stops the process, SB (as well as P) does worse, so it is in SB's interest—even as the M-powerful player who can out-last P—not to move from (3,4). Moreover, since P would prefer that the process stop at (2,3) rather than (4,2), he can threaten to stop it at (2,3) unless SB stops it at (3,4) first. Whether this threat is communicated or not, it is implicit in the moves and countermoves allowed by rules I–IV and SB's M-power; presumably, rational

Figure 5.3 *Outcome Matrices in Which M-Power Outcome Is Worse for SB or Not a Nash Equilibrium*

	Game 1 P			Game 2 P			Game 3 P	
SB	(3,4)* ←	(4,2)	SB	(2,3)* →	(3,2)	SB	(2,4) ←	(4,1)
	⇓	↑		↑	⇓		⇓	↑
	(2,3) →	(1,1)		(1,4) ←	(4,1)		(1,2) →	(3,3)*

Key: $(x,y) = (SB,P)$

4 = best; 3 = next best; 2 = next worst; 1 = worst

* = M-power outcome (whether SB or P has M-power)

Circled outcomes are Nash equilibria

Double arrows between outcomes indicate moves of player with M-power (assumed to be SB in this figure); single arrows, moves of player (P) without M -power

players would recognize this and, therefore, the soundness of making (3,4) the final outcome. Explicit threats, and the power they confer on SB in repeated play of sequential games, will be treated in Chapter 6.

Game 2 in Fig. 5.3, like the second Knowability Game in Fig. 2.2, is a game of total conflict with one Nash equilibrium, but it is structurally different from this game since only one player (P) has a dominant strategy. As the M-limited player, P, as in Game 1, can force SB to stop at an outcome that SB ranks lower [(2,3)]; this is better for P than the other outcome, (4,1), at which he can stop the process. Furthermore, SB cannot stop the process at (3,2), which he would prefer to (2,3), because P, by choosing his first strategy and not moving from it, can prevent a move to (3,2) in the first place.

If SB tried to depart from (2,3) to (1,4)—reversing the direction of the left double arrow in Game 2—the process would terminate at (1,4), SB's worst outcome, so SB would prefer that P stop it at (2,3). Thus, SB's M-power does not prevent implementation of an inferior outcome for himself (2), relative to P (3), in Game 2 in Fig. 5.3, as I showed earlier was true also for Game 1.

But M-power *alone* does not really hurt SB because, unlike the Revelation Game in Fig. 5.1 in which M-power matters, this attribute does not *change* the outcome: whichever player is M-powerful, the outcome is the same. Since players do not exercise effective power in these games, SB's inferior position, vis-à-vis P's, cannot, therefore, simply be chalked up to his M-power.

Moreover, the M-power outcome in these games, though not

the best for one (Game 1) or both (Game 2) players, is nevertheless Pareto-superior, unlike the Pareto-inferior (2,2) outcome induced in the Testing Game when SB is omniscient. However, the M-power outcome need not be a Nash equilibrium, as it is in Games 1 and 2. As a case in point, outcome (3,3) in Game 3 in Fig. 5.3 is next best for both players—neither player is worse off vis-à-vis the other—but not the Nash equilibrium, which is (2,4).

Yet (3,3), as well as all other M-power outcomes in games in which this outcome is the same whichever player is M-powerful, would appear relatively stable—natural places for the process to stop whichever player has M-power. These *M-neutral* outcomes coincide with what Hessel and I have called "absorbing outcomes," which we argue are the Pareto-superior outcomes that have the greatest attraction for *both* players in games without nonmyopic equilibria.[4]

5.4. *Theological Interpretations of Moving Power*

Because M-power is ineffective in the games in Fig. 5.3, which player is M-powerful is inconsequential. Thus, the (3,4) M-power outcome in Game 1 that hurts SB relative to P cannot be blamed on SB's lack of M-power—it is the same whether he or P is M-powerful. In the Relevation Game in Fig. 5.1, by comparison, SB can do better than the (3,4) outcome when he, rather than P, is M-powerful, inducing (4,2) instead.

But in the Revelation Game, as I have interpreted it, P is *not* the player with M-power, so (3,4) would not be the outcome implemented. On the other hand, if SB has this power, he can implement (4,2) and ultimately force P to believe even though SB does not reveal himself.

I say "ultimately," because the source of SB's power is his ability to continue moving. This means that he will switch back and forth between revelation and nonrevelation as P oscillates between belief and nonbelief. Eventually, SB's M-power would "come to rest" at (4,2)—nonrevelation by SB and belief by P—though one might question whether, after SB has switched from revelation to nonrevelation, he has not in fact revealed himself—once and for all—because of his prior choice.

4. Steven J. Brams and Marek P. Hessel, Absorbing outcomes in 2 × 2 games, *Behavioral Sci.* 27, 2 (October 1982), 393–401.

Plainly, the notion of "prior choice" needs clarification. If the "thunder and lightning, and a dense cloud . . . and a very loud blast of the horn" (Exod. 19:16) at Mount Sinai—before God "called Moses to the top of the mountain" (Exod. 19:20) to give him the ten commandments—constitutes revelation to readers of the Bible, and this is sufficient evidence for them today, then God cannot succeed in dropping out of sight. Others, on the other hand, may require a more immediate revelatory experience to believe in God's existence, and some indeed may find it. For those who do not, God presumably remains hidden or beyond belief unless they can apprehend Him in other ways.

The point is that, at least for some, a SB with M-power cannot inspire belief without, on occasion, revealing himself. Whether the occasion is reading the Bible for evidence of God's existence, or a personal revelatory experience, or something in between— such as regularly going to religious services—belief without some direct or indirect evidence of SB's presence seems difficult to support.

Doubtless, different people will seek out different kinds of evidence to sustain their beliefs in SB's existence. Should the sought-after evidence be found to be totally lacking, however, it is reasonable to suppose their beliefs will falter.

This last statement requires qualification. If some people are deluded into thinking, perhaps because of biased information they receive, that God exists (or does not), they will believe—incorrectly—that there is supporting evidence. It is their *perceptions* of the evidence, not the evidence itself (whatever that is), that counts.

Apart from the question of what constitutes evidence, the wavering between belief and nonbelief that SB's M-power induces for P in the Revelation Game suggests that one's belief in a SB has a *rational basis for being unstable*. Sometimes the evidence manifests itself, sometimes not. What is significant about the Revelation Game is that the rational moves of a M-powerful SB are consistent with SB's sporadic appearance, as he responds to belief in him by nonrevelation, to nonbelief by revelation.

This back-and-forth movement may extend over one's lifetime, as seeming evidence appears and disappears. For example, the indescribable tragedy of the Holocaust has shaken the faith of many believers, especially Jews, in a benevolent God, and for some it will never be restored. (This loss of faith is rooted in the problem of evil, mentioned in Section 2.4, to which I shall return in Chapter 6.)

On the other side, many nonbelievers have conversion experiences—sometimes induced by some mystical episode—and, as a result, pledge their lives to Christ or God. For still others, there is a more gradual drift toward or away from religion and belief in a SB, which is sometimes correlated with age. More broadly, there are periods of religious revival and decline, which extend over generations and even centuries, that may reflect a collective consciousness of the presence or absence of a SB—or maybe both: "The world manifests God and conceals Him at the same time."[5]

It is, of course, impossible to say whether a SB, behind the scenes, is planning his moves in response to the moves, in one direction or another, of individuals or society. But this is not the first Age of Reason, though it has had different names in the past (e.g., Age of Enlightenment). Nor will it be the last, probably again alternating with periods of religious reawakening (e.g., as occurred during the Crusades) that will continue to come and go. This ebb and flow is inherent in the instability of moves in the Revelation Game, even if a M-powerful SB in the end has his way and is able to implement (4,2).

The problem for SB is that peoples' memories fade after a prolonged period of nonrevelation, and the foundations that support belief crumble. Nonbelief sets up the need for some new revelatory experiences, sometimes embodied in a latter-day messiah, followed by a rise and then another collapse of faith.

If rational play in the Revelation Game is consistent with upturns and downturns in faith, SB's M-power in the Testing Game (Fig. 5.1) in inducing (3,3) (P does not prepare for a test, and SB does not test him) has more of an evening-out effect—it is ranked the same by both players. This seems a reasonable outcome for SB, because a true test of P's faith could be biased by his preparation or foreknowledge, which SB wants to avoid at all costs. (Recall that God became extremely upset with Adam and Eve when they ate from the tree of knowledge, because this enabled them to distinguish good from evil, right from wrong.) The (3,3) outcome in the Testing Game is also acceptable to P, who can avoid the hard work of preparation if he thinks he will not be tested.

To be sure, in the Bible God subjects some to severe testing. It is particularly gratifying for Him that when a righteous man like Job is caught off guard, he can pass a grueling test of faith, display-

5. Leszek Kolakowski, *Religion* (New York: Oxford University Press, 1982), p. 140.

ing great fortitude and courage when subjected to horrendous punishment.

I argued earlier that this is the best outcome for SB, the worst for P, in the Testing Game. However, if P survives the test—if he sustains his faith—as Job did, one might say that this (4,2) outcome becomes transformed into a mutually best (4,4) outcome.

This, in fact, seems to be an instance in which the fixed preferences of players in a game like the Testing Game fail to capture major changes that play of a game may produce in these preferences. The stability of the (3,3) outcome enforced by SB's M-power might also be challenged, for SB might occasionally have good reason to abandon it in order to test P before he has a chance to prepare.

The stability of (3,3) seems especially dubious in light of the fact that it is neither a Nash nor a nonmyopic equilibrium in the Testing Game, which means that this game lacks both short-run and long-run stability in the absence of M-power. Moreover, even if SB possesses this power, cycling may well occur before he is able to implement this outcome, giving play of the Testing Game the *appearance* of instability.

I conclude from this attempt to develop theological interpretations of M-power that the instability of P's choice of belief/nonbelief in the Revelation Game, and preparation/nonpreparation in the Testing Game, may be propelled toward more stable belief and nonpreparation by SB's M-power. But, having said this, I hasten to add that the essence of M-power is, of course, movement. This may benefit SB in the end because he is indefatigable, but the road to his M-powerful outcome may be strewn with a goodly number of strategy changes that both players make before SB's M-power is decisive. Even then, as I illustrated in Games 1 and 2 in Section 5.3 (Fig. 5.3), SB may have to cede the better outcome to P, depending on the game that is being played.

5.5. *Staying Power*[6]

In Section 5.2, I showed how the four sequential rules of play, based on the theory of moves and described in Section 4.4, could be modified to distinguish the player with M-power from the

6. Material in this and the next two sections is based largely on Steven J. Brams and Marek P. Hessel, Staying power in sequential games, *Theory and Decision* (forthcoming); see also D. Marc Kilgour, Equilibria for far-sighted players, *Theory and Decision* (forthcoming), who extends the notion of nonmyopic equilibrium (see citation in note 7) to most games in which S-power is not effective.

player without it in simple two-person games. In particular, by assuming a constraint on rule IV that allowed one but not the other player to continue moving indefinitely, an asymmetry giving one player M-power was defined.

Instead of assuming a constraint on the operation of rule IV, S-power (for staying power) modifies rules I and II to differentiate between the two players. Specifically, I assume that the player with S-power (S) chooses his strategy after the player without it ($\bar{S}$), so S knows what strategy $\bar{S}$ chose (as if he were omniscient and they made simultaneous strategy choices).

After $\bar{S}$ and S make their choices, an initial outcome is defined, and—continuing the alternation of moves—it is then $\bar{S}$'s turn to move if he chooses. These assumptions alter rules I and II but leave rules III and IV virtually intact, giving the following new rules of play when one player has S-power and the other does not:

I′. $\bar{S}$ chooses a strategy, followed by S, which defines an *initial outcome*.

II′. Once at an initial outcome, $\bar{S}$ can unilaterally switch his strategy and change that outcome to a subsequent outcome in the row or column in which the initial outcome lies.

III′. S can respond by unilaterally switching his strategy, thereby moving the game to a new outcome.

IV′. These strictly alternating moves continue until the player with the next move chooses not to switch his strategy. When this happens, the game terminates, and the outcome reached is the *final outcome*.

Note that rules I′–IV′, unlike rules I–IV, assume not only that the sequence of moves and countermoves from an initial outcome is strictly alternating but that the initial strategy choices are strictly alternating also, with $\bar{S}$ choosing first. In this sense, games played according to rules I′–IV′ are "more sequential" than games played according to rules I–IV, for even the initial strategies are chosen sequentially, not simultaneously. Henceforth, games played according to either set of rules will be considered sequential, though the two sets are distinct and are used to define different notions of omnipotence.

It is not obvious that S's ability to choose second confers on him any advantage, just as omniscience may lead to problems when choices are simultaneous. True, S's knowledge of $\bar{S}$'s initial strategy choice gives him additional information, but, by the same token, $\bar{S}$ would appear to be better able to cut off certain outcomes from the start by choosing first.

To illustrate this point, again consider Chicken in Fig. 5.2.

Now if SB is S, P would choose first according to rules I'–IV'. If he chooses his second strategy, SB would know this, and to avoid his as well as P's worst outcome, (1,1), it would appear that SB should then choose his first strategy, yielding (2,4), the best outcome for P but only the next-worst for SB. But this initial outcome would be the final outcome since P would have no incentive to move from it.

What are the consequences of SB's choice of his second strategy, instead of his first, when P chooses his second? This would yield (1,1) as an initial outcome, and P clearly would have an incentive to move from it to (4,2), as allowed by rule II'. But this gives P only his next-worst outcome, from which SB, obtaining his best outcome, would have no incentive to depart.

Recognizing this problem, assume that P chooses, according to rule I', his first strategy initially. Now it would appear that SB should choose his second strategy, yielding (4,2), but this choice would enable P to move next to (1,1), forcing SB to move to (2,4), which would then become the final outcome since P would have no reason to move further.

This leaves SB with only one remaining option—to choose his first strategy after P chooses his, yielding (3,3). Should P subsequently move to (2,4), SB can respond by moving to (1,1), forcing P to move to (4,2), which would then become the final outcome since SB, with the next move, would not return to the original (inferior) outcome, (3,3).

Because (3,3) is better than (2,4) for SB, SB will choose his first strategy when P chooses his. Since P would only do worse by choosing his second strategy, as already demonstrated, rational choices in sequential Chicken will result in (3,3) when played according to rules I'–IV'. Coincidentally, as shown in Section 4.4, (3,3) is the outcome that would be chosen if Chicken were played according to rules I–IV, for both players would, according to rule I, choose their first strategies *simultaneously*.

In sequential Chicken, because of the symmetry of the game, the rational choice of (3,3) is not affected by which player (SB or P) is S. In Section 5.6 I shall show that rules I–IV and I'–IV' may induce different outcomes, but whichever set of rules is operative, I again assume Nonmyopic Calculation (see Section 5.2). An algorithm for implementing this calculation when one player has S-power will be illustrated in Section 5.6.

Next, a specific criterion for ascertaining when moves from an initial outcome are rational is needed:

Rational Termination: S̄ will move from an initial outcome if and only if he can ensure a better outcome for himself before the process returns to the initial outcome and a new cycle commences. S will move (after S̄) from a subsequent outcome if he can either ensure a better outcome for himself before cycling or can force a return to the initial outcome through cycling. (The nonmyopic calculations implied by Rational Termination, which puts the onus on S̄ not to start moves that S can force to cycle back—should S *not* obtain his best outcome at the move preceding cycling—will be illustrated in Section 5.6.)

Implicit in Rational Termination is a presumed aversion on the part of the players to cycling. This is not to say that it may not occur in games; indeed, the M-powerful player's ability to continue to cycle is what I assumed to be the basis of his M-power. Here, however, I assume that players prefer the best determinate outcome they can ensure to cycling, given rules I'–IV' (which do not by themselves preclude cycling).

To be sure, cycling may be rational, but within the ordinal framework assumed here—wherein players can rank but not assign values to the outcomes—they would be unable to make this determination. Admittedly, players who can assign cardinal utilities to outcomes might derive greater expected utility from averaging over the four outcomes (constituting a cycle) to a guarantee, say, of only their next-worst outcome. Such a calculation, however, would require assumptions not only about what cardinal utilities are associated with each outcome, but also what the probabilities of being at each one are, along the lines developed for Newcomb's problem and Prisoners' Dilemma in Chapter 3.

But an expected-utility approach is different in other ways, too. In particular, I have assumed that the value that players derive from a game is realized only at a final outcome—initial strategy choices that produce an initial outcome, and subsequent moves and countermoves to other outcomes, contribute negligibly to the players' payoffs. It is the final outcome that counts, which is the best possible outcome that the players—thinking nonmyopically—can ensure that bars cycling, given rules I'–IV' and the provision of staying power for one of the players.

Rational Termination is somewhat different from the rationality criterion used to define nonmyopic equilibria according to rules I–IV, which essentially said that outcomes were nonmyopically stable only if the departure of either player from such an outcome would trigger a sequence of moves and countermoves that would result in an inferior final outcome. This necessary condition was rendered sufficient by a stopping rule to prevent cycling, which required that an outcome reached in the move-countermove process be best for the player who had the next move.

Rational Termination obviates the cycling problem by stipulating that a "better outcome" be found *before* the process returns to the initial outcome and then resumes. Otherwise, players will not depart from the initial outcome—only to return the process there—from which new departures would occur *ad infinitum.* Hence, no arbitrary stopping rule is required because all cases— with or without possible cycling—are covered by Rational Termination.

When coupled with Nonmyopic Calculation, the move-countermove consequences of all initial strategy choices can be determined, whether SB or P is the S player. Both players can then select their strategies that lead to the final outcome that is best for them, as the earlier informal analysis of Chicken illustrated. An algorithm for making these calculations in a formal and systematic way will be described and illustrated in Section 5.6.

5.6. An Algorithm for Determining S-Power Outcomes

So far I have used Chicken to show, in a somewhat informal manner, why $\bar{S}$ and S would be well advised to choose (3,3) initially and then remain there: it is the final outcome according to rules I' –IV'. (As I shall show later, the initial outcome that rational players choose may differ from the final outcome.) But still lacking is a systematic procedure for finding both the rational (final) outcome and the rational (initial) strategy choices of $\bar{S}$ and S.

Again consider Prisoners' Dilemma, shown in Fig. 5.4. Assume, as before, that P is $\bar{S}$ and SB is S. Their possible strategy choices, and subsequent moves, according to rules I'–IV', are shown in the game tree in Fig. 5.4. Thus, starting at the top of the tree, P can choose either his column 1 or column 2 strategy, and SB, knowing this choice, can choose either his row 1 or row 2 strategy.

Depending on which choices P and SB make, one of the four possible outcomes, shown in Fig. 5.4, will occur. Then, from each of these outcomes, P and SB can move and countermove if they wish.

There is not space in Fig. 5.4 to list all possible moves and countermoves from each of the four outcomes, so I have listed only the tree of possible choices emanating from outcome (4,1).

Figure 5.4 *S-Power Outcome of Prisoners' Dilemma*

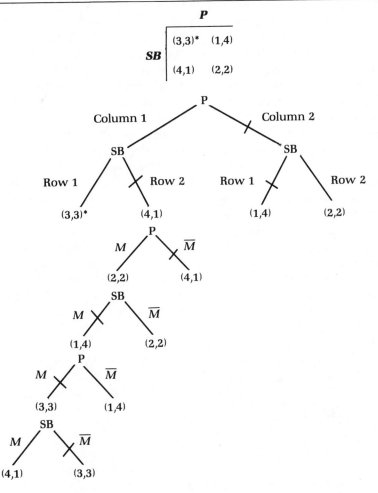

Key: $(x,y) = (SB,P)$
 4 = best; 3 = next best; 2 = next worst; 1 = worst
 M = move; $\overline{M}$ = don't move
 * = S-power outcome

Thus, P can move or not move to (2,2), whence SB can move or not move [if P moved to (2,2)] to (1,4), whence P can move or not move [if SB moved to (1,4)] to (3,3), and finally SB can complete the cycle by returning the process to the initial outcome, (4,1).

To determine which moves will be made by a rational

player, using the criterion of Rational Termination, start at the bottom of the tree and work backward. (The argument made here is essentially the same as that made for Chicken in Section 4.4 and Prisoners' Dilemma in Section 4.5, but since the game tree is now more complex, the details seem worth providing.) Starting at the bottom, SB (as S) would move from (3,3) to (4,1) by Rational Termination, since (3,3) is not his best outcome, so slash the don't-move branch to (3,3), making (4,1) the surving outcome. Comparing this outcome with (1,4) at the next higher level, P would not move to (3,3) [and then (4,1)], so slash the move branch connecting (1,4) to (3,3) to indicate this move would not be made

At the next higher level, SB would compare the surviving outcome (1,4) with (2,2); because he would prefer (2,2), slash the move branch connecting (2,2) and (1,4). Now, at the next higher level, P would compare (2,2) to (4,1), and since he would prefer (2,2), slash the don't-move branch connecting (2,2) and (4,1).

This leaves (2,2) as the surviving outcome of the move-countermove process that would commence at (4,1). That is, it would be rational for P to move from (4,1) to (2,2), but then the process would terminate. In contrast, it is not rational for P to move from any of the three other outcomes, as can easily be demonstrated by a similar backward-induction analysis from each of these outcomes.

What remains is to analyze the top part of the game tree describing the players' initial strategy choices, beginning with SB, the second-moving player with S-power. Since SB would prefer row 1 if P chose column 1, and row 2 if P chose column 2, I have slashed his row branches leading to (4,1) and (1,4).

With these outcomes eliminated, P faces the choice, at the top of the tree, between (3,3) and (2,2); he would obviously prefer (3,3), so slash the column 2 branch. The unslashed branches that remain indicate that both P and SB would choose their first strategies initially, yielding (3,3), from which P would have no incentive to move. Thus, (3,3) emerges at the S-power outcome of Prisoners' Dilemma when played according to rules I'–IV'.

It is not difficult to show that if the game were played according to rules I–IV, (2,2) as well as (3,3) would be nonmyopically stable—the players would not have an incentive to move from either outcome if it were chosen initially. The reason is that without a specified order of player choice (e.g, first P, then SB, as in Fig. 5.4), one cannot in general ascertain what strategy choices the players will make initially, and hence what the initial outcome will be [though the two-sided analysis of Prisoners' Dilemma discussed in Section 4.5 would single out (3,3)].

Despite this indeterminacy, one can assess the stability of every outcome, given that it is selected. As indicated in Fig. 5.4, P would not move from either (3,3) or (2,2), and neither would SB by the symmetry of the game, so these outcomes are nonmyopically stable. On the other hand, SB, who as S cannot move initally from (4,1) according to rules I'–IV', would move under rules I–IV to ensure (3,3) rather than (2,2), so (4,1) is not a nonmyopic equilibrium.

In this manner, rules I'–IV' single out (3,3) as the S-power outcome in sequential Prisoners' Dilemma, whereas rules I–IV show (2,2) also to be nonmyopically stable. With the sole exception of sequential Prisoners' Dilemma, which is the only 2 × 2 game with more than one nonmyopic equilibrium, the nonmyopic equilibria in all 37 games that have them are the rational outcomes according to rules I'–IV'. Sequential Prisoners' Dilemma is the exception because it has two nonmyopic equilibria, and perforce one is eliminated by the game-tree analysis under rules I'–IV'. When preferences are strict, these rules always induce a (single) determinate outcome for the binary comparisons at each stage, and, finally, permit only one outcome to survive at the top of the tree.

Fortunately for the players, it is the cooperative outcome, (3,3), rather than the noncooperative outcome, (2,2), that Rational Termination finds. [It should be recalled that the different argument given in Section 4.5 for eliminating (2,2) in Prisoners' Dilemma does not require Rational Termination.] In sequential Chicken, too, this rationality sustains (3,3), though the cooperative outcome in neither game, when played nonsequentially, is a Nash equilibrium and hence stable in a myopic sense. In fact, the cooperative (3,3) outcomes in sequential Prisoners' Dilemma and Chicken are the *only* nonmyopic equilibria that do not coincide with Nash equilibria.[7]

The game-tree analysis of Prisoners' Dilemma illustrates how S-power outcomes can be systematically determined by working up from the bottom of a game tree and eliminating inferior choices. Recall that Rational Termination requires that $\bar{S}$ be able to ensure a better outcome before the process returns to the initial outcome, and that S, *if put in the position*, will force a return to the initial outcome should he not obtain his best outcome at the move preceding cycling. Thus, it is $\bar{S}$'s sole responsibility to prevent cycling and retrogression to "square one."

In the case of Prisoners' Dilemma, self-restraint by $\bar{S}$ is unnec-

7. Steven J. Brams and Donald Wittman, Nonmyopic equilibria in 2 × 2 games, *Conflict Management and Peace Sci.* 6, 1 (1983).

essary. If (3,3) is the initial outcome, an (irrational) departure by either player (as $\bar{S}$) would move the process to (2,2). There it would stop well before cycling could begin, as shown by the game tree in Fig. 5.4, commencing with outcome (4,1)—caused by a departure by SB from (3,3)—and terminating at (2,2). In the case of Chicken, if P is $\bar{S}$, his departure from (3,3) as the column player in Fig. 5.2 would trigger a sequence of moves that would eventually lead to (4,2), from which SB could move the process to (3,3) (if there were no Rational Termination) but would have no incentive to do so since (4,2) is his best outcome.

Now consider the Revelation Game in Fig. 5.5 (cf. its M-power representation in Fig. 5.1), whose game tree when P is $\bar{S}$ is shown below the outcome matrix. Note that (4,2) is the S-power outcome whichever strategy (first or second) SB chooses initially. The reasoning is as follows:

1. On the left-hand side of the tree, (3,4) will not be chosen because SB cannot better (4,2), where the process will stop if P chooses column one initially. This is so because if P moved to (2,3) and SB to (1,1), P would have to stop the process at (1,1) in order not to violate Rational Termination, for moving to (3,4) would induce cycling back to (4,2). Since (1,1) is worse for P (as well as SB) than (4,2), P would not move to (2,3) from (4,2).

2. On the right-hand side of the tree, (2,3) will not be chosen because if it were, P would not move to (4,2) since SB would terminate the process there. On the other hand, if (1,1) were the initial outcome, the game-tree analysis in Fig. 5.5 demonstrates that it is rational for both players to move the process to (4,2). For if P subsequently moved the process to (2,3), SB would move it back to the initial outcome, (1,1), and P would thereby have violated—in going from (4,2) to (2,3)—Rational Termination: he would *not* have ensured a better outcome than the initial outcome, (1,1), before the process cycles back to this outcome.

To sum up, the algorithm for finding S-power outcomes involves a game-tree analysis modified by a stipulation that—consistent with Rational Termination—says that $\bar{S}$ will consider only moves that cannot lead to cycling. Thus, moves to the level before cycling by $\bar{S}$ are technically forbidden, for they could put S in the position of completing the cycle, which I assumed he would always do except when he obtains his best outcome (4). This is not the case for SB at the bottom of the Fig. 5.5 tree.

To be sure, it may not be rational for S to complete the cycle, as in Prisoners' Dilemma and Chicken starting from initial outcome (3,3), but in games like the Revelation Game, it may be of

Figure 5.5 *S-Power Outcome of Revelation Game When SB Is S (Player with Staying Power)*

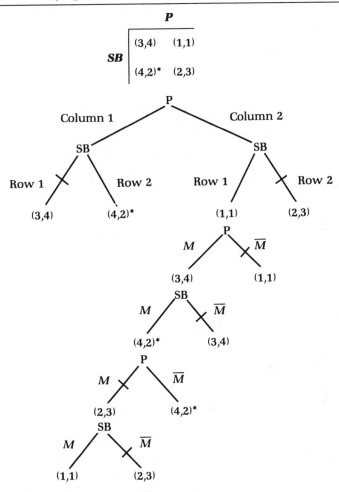

Key: $(x,y) = (SB,P)$
4 = best; 3 = next best; 2 = next worst; 1 = worst
M = move; $\overline{M}$ = don't move
* = S-power outcome when SB is S

immediate benefit for SB (as S), if put in the position of being able to return the process to the initial outcome, to do so. Thus, if (4,2) were the initial outcome in the Revelation Game, and P (irrationally) moved to (2,3), inducing SB to move to (1,1) and causing P to move subsequently to (3,4), it is reasonable to suppose that SB would complete the cycle. Not only would SB immediately benefit at (4,2), but P would have violated Rational Termination in moving from initial outcome (4,2) without ensuring a better outcome for himself before the process returns to (4,2).

It can be shown that Rational Termination, in forbidding cycling, prevents Pareto-inferior outcomes, like (1,1) and (2,3) in the Revelation Game, from being implemented.[8] It should be stressed, however, that the ban on cycling emerges from individualistic rational calculations, not from a consensus that the players somehow arrive at cooperatively: to "ensure a better outcome for himself" under Rational Termination means to do so by oneself without help from S, who will always move except under the conditions specified by Rational Termination.

This essentially noncooperative view of game playing is exactly why $\bar{S}$ (P) must take the precaution of not putting S (SB) in a position to complete the cycle. Such a precaution becomes especially prudent when SB can gain from cycling, either immediately or perhaps from averaging over all outcomes if cardinal utilities are assumed.

It turns out that the S-power outcomes are the same as the M-power outcomes in all games discussed so far in this chapter. Thus, for the three games in Fig. 5.3, S-power, like M-power, does not enable SB to obtain his best outcome, or even one ranked as high as P's in Games 1 and 2. SB can ensure (3,3) in Game 3, as well as in Chicken, Prisoners' Dilemma (both shown in Fig. 5.2), and the Testing Game (Fig. 5.1). In the Revelation Game (Fig. 5.5), as I have just shown, S-power for SB does in fact lead to his best outcome, (4,2), whereas P's best outcome, (3,4), would result if he had S-power.

This congruence of M- and S-power outcomes means that the theological implications of M-power in the Revelation and Testing Games, discussed in Section 5.4, carry over to S-power in these games. In the case of S-power, however, SB's ability to secure the M-power/S-power outcomes in these games stems from his ability to hold out until P makes his initial strategy choice, after which the players respond to each other sequentially but have an aversion to cycling.

There are differences, however, in the effects of moving and staying power. I shall illustrate these differences in Section 5.7, beginning with an example that has some theological significance.

---

8. This is in fact true of S-power outcomes in all two-person finite games, as shown in Brams and Hessel, Staying power in sequential games.

5.7. *M-Power Versus S-Power in the Commitment Game and Other Situations*

Assume that P has two strategies:

1. Make a commitment to SB
2. Await a commitment from SB

SB has two analogous strategies:

1. Make a commitment to P
2. Await a commitment from P

Assume that the players have the same goals:

(1) Primary goal—that one player make a commitment, the other player wait
(2) Secondary goal—if primary goal satisfied, be player that waits; if not, prefer that both players commit rather than wait

The game defined by these strategies and goals, which I call the Commitment Game, is shown in Fig. 5.6.

Like Chicken, it contains two Nash equilibria, but neither player has a dominant strategy. Unlike Chicken, the Commitment Game does not have a nonmyopic equilibrium [(3,3) in Chicken]

Figure 5.6 *Outcome Matrix of Commitment Game*

		P	
		Make commitment	**Await commitment**
	Make commitment	Neither waits (2,2)	P waits ⃝(3,4)
SB			↕
	Await commitment	SB waits ⃝(4,3)* ⇆	Both wait (1,1)

Key: (x,y) = (SB,P)
4 = best; 3 = next best; 2 = next worst; 1 = worst
* = S-power outcome when SB is S
Circled outcomes are Nash equilibria
Bidirectional arrows indicate rational moves from initial outcomes (3,4) and (4,3) to other outcome

which, if it existed, would be the M-power outcome, whether SB or P had M-power.

In fact, as the bidirectional arrows in this game indicate, the player receiving his next-best outcome [say, SB at (3,4)] would be tempted to move the process to (1,1), whence the other player (P) would move it to his next-best outcome [(4,3)], which is best for the player (SB) who departed in the first place. But if the initial outcome were (4,3), P would reverse the process, as indicated by the arrows in the other direction.

In other words, there would be no cycling in the Commitment Game, according to rules I–IV. If the rules of play permitted the players to backtrack, however, the process would oscillate between (3,4) and (4,3) indefinitely. One could define the player with M-power to be the one who can continue the oscillation after the other player is forced to stop, which would enable him to realize his best outcome in the Commitment Game.

This new definition of M-power, in a noncyclical game like the Commitment Game, poses some difficulties of interpretation. Is it reasonable to characterize SB as M-powerful if he can go back and forth on his commitment longer than P can? What meaning does a "commitment" have that is continually broken and then reinstated?

A commitment in religion traditionally means a vow that cannot easily be retracted. Christian theology, by and large, places the onus on human actors; at least since Saint Augustine, it has been expected that they pledge themselves to God or Christ.[9]

Judaism does not reverse this burden entirely, but the premise of a "chosen people" is that God is significantly involved in making the choice. The sacred covenant that He establishes with Israel is two-sided, and, at least in principle, rewarding to both parties, though the Bible reports interminable conflict between the parties. To come full circle, certain Christian theologians, such as Karl Barth, have argued the insignificance of human choice—only God can decide to approach us, never vice versa.[10]

Whoever takes the initiative in the Commitment Game, I assume he does less well, comparatively speaking, than the player who waits *and* receives the other player's commitment [(3,4) and

9. Saint Aurelius Augustinus, *The Confessions of Saint Augustine*, translated by Edward B. Pusey (New York: Modern Library, 1949).

10. Karl Barth, *Church Dogmatics*, translated by G. W. Bromily (New York: Harper & Row, 1962); see also Hans Küng, *Justification: The Doctrine of Karl Barth and a Catholic Reflection* (Philadelphia: Westminister, 1981).

(4,3) in Fig. 5.6]. In this situation, wherein one player initiates and the other responds, the initiator does better (3) than when both commit simultaneously [(2,2)], because I assume there will be problems in aligning what are, perhaps, the different expectations related to the two players' commitments to each other. [The coordination problem might not exist—commitments might mean the same to both players—but the Bible is filled with descriptions of misunderstandings of appropriate offerings and sacrifices to God (e.g., by Cain and Saul) that suggest that this is not an unreasonable assumption to make.] Still, this is a better outcome for both players than their playing some kind of waiting game, each deferring to the other [(1,1)], and thereby never succeeding in establishing any kind of satisfactory relationship.

The player's preferences in the Commitment Game, and the previously postulated goals underlying them, are, of course, arbitrary. The different theological assumptions I alluded to earlier about who bears responsibility for choosing whom undoubtedly make no "commitment game" descriptive of all religious faiths, or even to all members of one faith.

One change of preferences in the Commitment Game does not matter, however. If the orderings of the players for the two best outcomes were reversed—(3,4) were substituted for (4,3), and (4,3) for (3,4), in the Fig. 5.6 outcome matrix—the conclusions I shall draw presently about the effects of S-power would not be altered.

The basic result is that if SB has S-power in the Commitment Game, he can ensure (4,3), his best outcome. The reason is that if P must act first, either of his strategy choices will lead to (4,3):

1. If P makes a commitment, so will SB, resulting in (2,2), from which rational moves and countermoves by P and SB will take the process to (4,3), where it will stay.
2. If P awaits a commitment, so will SB, resulting in (1,1), from which a rational move by P will take the process to (4,3), where it will stay.

Curiously enough, in each case the players start out at a Pareto-inferior outcome (for similar moves in overcoming the paradox of omniscience, see Section 4.4) before SB's S-power enables him to implement (4,3). The reason that SB eschews both (3,4) and (4,3) initially is that P would not move from the former outcome (his best), whereas P would move from the latter outcome (his next-best) to induce (3,4), his best outcome.

One lesson of rational moves in this game is that SB's S-power gives him the ability to select an initial outcome that is immediately disadvantageous, but ultimately advantageous, because it creates an opportunity for him eventually to be the player who awaits P's commitment. But if he actually prefers to make the initial commitment, and (3,4) and (4,3) are interchanged in the Fig. 5.6 matrix, he would still be able to ensure his best outcome with S-power, whichever strategy P chooses initially.

S-power, as it takes shape in rule I', can also be interpreted as omniscience, because it allows SB to respond to P's initial strategy choice. If no moves and countermoves from the initial outcome are allowed, there is a paradox of omniscience in the Commitment Game (see Section 4.2): P, knowing that SB can predict his "await" strategy, can force SB to make a commitment, leading to (3,4), P's best outcome and SB's next-best.

The fact that SB, by virtue of his S-power, can position himself to achieve his best outcome eventually might seem contrived. After all, do people really strategize about their commitments, making them and withdrawing them because it is strategically advantageous to do so in a game with SB? And is it reasonable to suppose that SB is simply a clever strategist?

I believe the larger question that underlies play of the Commitment Game concerns, as in the Revelation and Testing Games, the instability of beliefs, expectations, and, in the present game, commitments. Now the Commitment Game does not cycle, as do the other two games, so SB's M-power is inconsequential (unless it is redefined, as previously suggested, to allow him to continue the oscillation process longer than P). If backtracking is not permitted, however, only SB's S-power "works" in this game; furthermore, SB's using it to secure his best outcome is roundabout, as I have shown, involving some maneuvering by the players at the start.

This is perhaps the crucial point. Since belief in SB may need reinforcement from time to time, perhaps by revelation, occasionally SB may find it necessary to take the initiative when P's own commitment lags. Ultimately, as I assumed, SB would prefer that P take the initiative, but if there is a lapse on P's part, it may be rational for SB to step in—at least temporarily—until he can swing the pendulum in the other direction. In the end, SB's S-power, but not M-power, can set the process back on course (SB's) in the Commitment Game, which illustrates the limitations of M-power as I have defined it.

But S-power has limitations, too, illustrated by the game in Fig. 5.7. In this case, SB's M-power enables him to implement an

Figure 5.7 *Outcome Matrix of Game in Which M-Power Is More Potent Than S-Power for SB*

P

(2,4)^S	(4,1)
(1,2)	(3,3)^M

SB

Key: $(x,y) = (SB,P)$
4 = best; 3 = next best; 2 = next worst; 1 = worst
Circled outcome is a Nash equilibrium
Superscripted outcomes indicate those SB can implement if he has M-power and S-power

outcome preferable to his S-power outcome, as shown. This is, in fact, the only 2 × 2 ordinal game in which SB's M-power overrides his S-power, whereas the five games in Fig. 4.4, which include the Commitment Game, are the only games in which SB's S-power induces his best outcome but his M-power has no effect. M-power has no impact because, though it is rational for the players to move from one Nash equilibrium to the other (as in the Commitment Game), they would not cycle over the four outcomes, which is necessary for the exercise of M-power by one player.

The potency of M-power versus S-power, then, is decidedly mixed. In most games they induce the same outcomes, but in a few, as I have shown, one kind of power is more potent than the other. Indeed, in the game in Fig. 5.7, SB gains no advantage from possessing S-power, because having to move second does not allow him to implement his best outcome but does allow P to implement his.

The "failure" of M-power in this game is not unlike the "failures" of M-power we found in Games 1 and 2 in Fig. 5.3, in which SB's M-power led to an outcome that he ranked lower than P. Nevertheless, S-power, like M-power, always leads to Pareto-superior outcomes.

There is, however, one major difference between M-power and S-power: when S-power is effective, leading to different outcomes depending on which player possesses it, it may be so in a perverse way. Specifically, in the game in Fig. 5.7, if P is the player with S-power, he can induce (3,3), the same outcome that SB's M-power induces. Paradoxically, this outcome is better for SB, whereas the S-power outcome for SB, (2,4), is better for P.

Thus, it is in the interest of each player that the other have S-power in the game in Fig. 5.7, which casts doubt on whether S-power captures a valid notion of power, at least in this game. However, in the other 19 games in which S-power is effective, it favors the S-powerful player (i.e., he can do better when he has S-power than when the other player has it), so, generally, it is desirable for a player to possess S-power. This one exception, though, certainly points to a "paradox of S-power," which, as I indicated in Section 5.3, does not characterize effective M-power, even though the M-powerful player may do worse than the M-limited player when M-power is not effective.

5.8. Conclusions

As the definition and measurement of power have bedeviled political scientists, so theologians and philosophers have struggled with the meaning of omnipotence and conditions that govern its exercise by God or some other superior being. In this chapter my view has been that total control by a SB over all decisions and actions of P would subvert his free will.[11]

This view, in my opinion, is compatible with the exercise of power by God in the Bible, a central tenet of which is that man is generally free to do what he pleases. To be sure, his choices have consequences, as they do in everyday life, and rather frequently they do not comport with what God thinks is right and proper. This, incidentally, does not seem due to God's lack of trying to guide the people of the Bible.

Throughout this book, I have followed the Bible in assuming that P can make independent strategic choices. Without violating P's free will and sovereignty, I think it appropriate to endow SB with certain powers in a game but not to make him so strong that he alone can dictate the outcome. To endow SB with the ability to stop P from moving, for example, even when it is rational for P to do so, seems inconsistent with allowing P to make rational choices

---

11. It would also imply a kind of "metaphysical determinism," which Karl R. Popper rejects in *The Open Universe: An Argument for Indeterminism*, ed. W. W. Bartley III (Totowa, NJ: Rowman and Littlefield, 1982), pp 87–92.

and be subject to essentially the same rules that govern SB's actions (i.e., permit him to move and countermove like SB).

Within the framework of the theory of moves spelled out in Chapter 4, M-power and S-power seem to be attributes that SB might possess and use to promote his ends. Their intuitive meaning can be given formal definition in a game, as I have shown, and consequences can be derived from the exercise of each kind of power in sequential games.

In the Revelation Game, SB can induce belief without revelation, and in the Testing Game, he can provide an escape from the Pareto-inferior (2,2) outcome that his omniscience induces, with either kind of power. Similarly, in Chicken and Prisoners' Dilemma, both M-power and S-power enable SB to implement the compromise (3,3) outcome, though in some games these different kinds of power lead to higher-ranked outcomes for the nonomnipotent player. In the Commitment Game, only staying power allows SB to implement his best outcome, though in the game in Fig. 5.7, moving power is more potent. Paradoxically, in this game, S-power is effective, but it hurts the S-powerful player and helps the S-limited player. This is never the case with M-power in games in which it is effective.

What seems common to all the theological games discussed in this chapter is an underlying instability that SB can exploit with adroit moves backed up by the power he possesses. This instability seems to mirror the ups and downs both individuals and society face over time in their beliefs about, and commitments to, a SB. In the next chapter, I shall add new qualities to the list of SB's supernatural attributes and derive their implications for the game outcomes they enable SB to implement.

SIX

Immortality and Incomprehensibility

6.1. Introduction

The moving and staying power that, as I suggested in Chapter 5, may distinguish SB from P can also be used to differentiate more powerful from less powerful actors in the secular world. There is nothing sacrosanct about these attributes, though I think that the indefatigability required of a player with M-power, and the suspension of choice required of a player with S-power, may well characterize aspects of omnipotence that a supernatural figure, who embodies the sacred and mysterious in a religion, may possess.

What no natural figure can lay claim to is immortality. Great heroes, whether pagan or religious, may pass on traditions, give rise to movements of schools of thought, or be remembered as martyrs or worshipped as saints. While memories of them may endure, however, they do not. Every natural creature dies.

God, the devil, and other incarnations of everlasting good and everlasting evil are not natural. (There seem to be no morally neutral supernatural figures, profane or sacred, who live forever.) For the purposes of this analysis, I shall not try to offer game-theoretic definitions of immortality, as I did for omniscience and omnipotence, but rather look at what I think is an important *correlate* of immortality, namely, the ability of an immortal player to make credible threats in repeated games.

By virtue of continuing as a player over time immemorial, an immortal player can threaten mortal players, and if necessary carry out his threats (to everybody's dissatisfaction, I assume, in the immediate game) without besmirching his rationality. For even if he hurts himself in carrying out such a threat, it is rational to do so if he calculates that it will deter mortal players in future games from ignoring his warnings.

In addition to being immortal, SB may be, or wish to be, incomprehensible. Recall that in Section 2.4 I discussed a series of Knowability Games in which SB chooses between appearing knowable or not. In this chapter I shall show why SB may wish to appear incomprehensible, either by revealing nothing about his preferences in a game or by randomizing his choices in a particular manner.

Hiding his preferences permits SB to make false announcements about them in order to induce P to make choices favorable to himself (i.e., SB). But sometimes it may be advantageous for SB not to project a knowability that is in fact specious but instead to be truly unpredictable. I shall demonstrate in a particular case that, with only partial omniscience, he can further his aims by choosing his strategies randomly so as to befuddle P. This essentially arbitrary behavior calls to mind, once again, the problem of evil.

6.2.   Threats and Immortality[1]

There were two trees in the garden of Eden, one of knowledge of good and evil and the other of life. After Adam and Eve defied God and ate from the former, God punished them and also made clear why the knowledge so gained from eating the fruit of this tree presented a problem:

> Now that man has become like one of us, knowing good and bad, what if he should stretch out his hand and take also from the tree of life and eat, and live forever! *(Gen. 3:22)*

1. Material in this and the next section is drawn from Steven J. Brams and Marek P. Hessel, Threat power in sequential games (mimeographed, 1982).

In other words, man, having become divine in his ability to distinguish good from evil, as the wily serpent had predicted, could seriously threaten God's authority if he also gained immortality from eating from the tree of life. It was, apparently, the combination of immortality and knowledge that God considered intolerable, so he banished the couple from the garden to arrest the immortality of an already knowledgeable Adam and Eve and thereby ensure His own unique and privileged position.

Setting aside the natural fear of death, why is immortality to be so cherished? Alone—without knowledge of good and evil—it will be simply an uninformed and, presumably, impotent presence. With moral knowledge, though, man can make ethical choices and, without death to confront, never have to worry about the ethics he adopts. God would lose his superiority as man becomes divine like Him.

God arrests man's moral freedom (abandon?) by making him mortal. It is, I believe, not that the God of the Old Testament fears man's immortality as such but rather that He feels He would lose control if man did not have to be concerned about his future because it is endless—the game would literally go on forever.

In previous chapters I postulated the single play of a game: while allowing sequential moves and countermoves by the players after they make their initial strategy choices, games were not assumed to be repeated. Now I want to investigate some implications of repeated play while retaining the sequential rules.

In particular, repeated play allows players to make threats, which, once carried out, are rendered credible in later games. Given that there is the wherewithal to back them up, which I shall call T-power (for threat power), one can analyze both the stability of outcomes in a repeated context and the effectiveness of threats.

Clearly, the ability of players to "stretch out" their calculations in anticipation of playing future games may destabilize outcomes stable in single play, just as myopically stable outcomes—such as Nash equilibria—are often rendered unstable when the rules allow moves and countermoves from initial outcomes. Similarly, outcomes unstable in the single play may be rendered stable when the game is repeated, and players anticipate its recurrence, which alters their rationality calculations in ways to be made precise later.

Games that comprise repeated plays of nonsequential games, in which only rule I (Section 4.4) is operative, are referred to as *supergames.* Although they have been extensively analyzed,[2] there has been no supergame analysis under the rules of sequential play. Yet, play under these rules—at least in the secular world—seems ubiquitous, and repeated play under these rules may be just as common. The willingness of parties, for example, to accept prolonged stalemates, to refuse to negotiate at all, or to resort to the use of force—all at a considerable cost—can often be explained only by their expectations of possibly having to face the same situation over and over again and trying to set a precedent of implacable firmness to deter future actions of opponents.

Before transporting this idea to the realm of superior beings, consider, as a more commonplace illustration of this kind of calculation, its application to international politics. Assume one superpower anticipates it will be continually engaged in the same kind of dispute with many small countries, but each small country anticipates it will have only one encounter of this kind with the superpower. If this is the case, the superpower must worry about its bargaining reputation, whereas the small country has no such concern because its stake is not continuing.

Call the continuing player (superpower) the *threatener,* and the noncontinuing player (small country) the *threatenee.* Then the threatener can make his threat credible by ignoring what he would lose in the short run if he were forced to carry it out (I assume there is always a cost to both the threatener and threatenee associated with the former's carrying out a threat) and instead focusing on the long-run value that a carried-out threat would have in enhancing the credibility of the threatener's *future* threats in repeated plays of the game.

It should be noted that "repeated plays of the game" may involve the threatener against either the same or different threatenees (as in the superpower–small-country example). Whether the threatenees are the same or different in repeated plays, I assume that they know that the threatener is willing and able to carry out his threats to establish his future credibility.

Insofar as the threatener establishes his credibility by carrying out his threats, it will plausibly extend to repeated plays of *different* games. Thus, a threatener who has built up a reputation for "toughness" by taking reprisals against opponents who have not

2. References to this literature are given in Steven J. Brams and Donald Wittman, Nonmyopic equilibria in 2 × 2 games, *Conflict Management Peace Sci.* 6, 1 (1983).

met his terms can be assumed to be a qualitatively different player from one who is out to ensure his best possible outcome in any single play of a game under the sequential rules of play. Similarly, the threatenee, aware of the threatener's reputation, will also be a different player.

Stated differently, each player will have goals different from those previously assumed in the single play of a sequential game. Before specifying each player's rational calculus, given the assumption of repeated play for one player and his concomitant concern for his credibility in future games, I first want to indicate the tie-in of threats to immortality.

Everyone worries about the future. One's present behavior inevitably affects how others see one and treat one later. So does the behavior of collectivities, such as organizations and nation-states, as I have just argued.

There is, nonetheless, a qualitative difference, I think, between looking ahead to the near versus distant future, especially when "distant" is measured in eons, as it is for an immortal player. Then one's image and reputation become of paramount concern because the time horizon is infinite. The immediate pales in significance; the eternity that lies ahead is what counts, swamping whatever payoffs accrue in any finite series of preceding games—given, of course, that one's reputation is kept alive.

An immortal player, therefore, plays for the future. If he must suffer present losses to deter untoward actions by future opponents, this is worthwhile, because his reputation—if sufficiently frightening—will bring at least some to their senses. Thus, in the Bible, the prostitute Rahab hides the Israelite spies in Jericho because

> we have heard how the LORD dried up the waters of the Sea of Reeds [Red Sea] for you when you left Egypt, and what you did to Sihon and Og, the two Amorite kings across the Jordan, whom you doomed. When we heard about it, we lost heart, and no man had any more spirit left because of you; for the LORD your God is the only God in heaven above and on earth below. *(Josh. 2:10-11)*

More crassly, because the conquering Israelites, with God on their side, are likely to prevail against her king, Rahab calculates that it would be prudent to strike a deal with them to save herself and her family. This she does and, true to her calculation, is rescued from the destruction of Jericho.

I have characterized the God of the Old Testament as someone with an

> overweening concern for His reputation [who] . . . continually
> broods about it. He worries endlessly about how to enhance it. He
> is not so much concerned with the world as how He thinks the
> world sees Him. He is other-directed with a vengeance.

> . . . As a consequence of His immortality, God's actions have both
> an immediate effect and a resonant effect on future generations.
> Knowing this, He must choreograph His moves with an eye
> toward the image they will convey to potential players, perhaps
> not yet born, in games that are hard even to envision.

> This explains, I believe, why God is so concerned about His image.
> He as much wants to impress future adversaries—and those of us
> who read of His exploits in the Bible—as be a rational player in the
> here and now. God quite candidly confesses this motive on several
> occasions, citing on the one hand His beneficient treatment of
> Abraham, Isaac, and Jacob, and on the other His harsh treatment
> of Pharaoh. The good prosper, the bad succumb.

> . . . He is a constant worrier who believes He must cement His
> image as a vigorous disciplinarian. This, He thinks, will forestall
> later deviations, and indeed it does make some people like Rahab
> and the Gibeonites more God-fearing.[3]

God's attitude and behavior, in my opinion, befit an immortal
player whose threats, and their perception by the people, often
have a decisive influence on the outcomes He is able to achieve in
biblical games.

Next I turn to a more formal analysis of threats in 2×2 ordinal games that are repeated, where I shall describe the conditions
that make threats credible and state some general propositions
about "compellent" and "deterrent" threats that are also applicable to larger games. I shall not couch this analysis in theological
terms, except to refer to the threatener as SB, but shall return in
Section 6.4 to a discussion of the applicability of this analysis to
theology, and particularly a game I call the Punishment Game.

6.3. *Repeated Plays*

Assume that one of the two players involved in a game, played
according to rules I–IV (Section 4.4), faces repeated plays of the

---

3. Steven J. Brams, *Biblical Games: A Strategic Analysis of Stories in the Old
Testament* (Cambridge, MA: MIT Press, 1980), pp. 173–176.

game. Call this player T (for threatener). The other player, $\overline{T}$ (the threatenee), is concerned with only a single play of the game; he is, however, aware of T's continuing involvement in the game.

As noted earlier, recurrence of a game may make it rational for T to threaten $\overline{T}$, and carry out this threat, even if it results in a worse payoff for both players when the game is played only once. To analyze the effects of repeated plays of a game on the stability of the game's outcome, I make the following assumption about T's *threat behavior:*

> T will stay at, or move to, a strategy disadvantageous to himself (as well as $\overline{T}$) in a single play of a repeated game if and only if it enables him to establish the credibility of his threats in future plays of games that ultimately will lead to better outcomes for himself.

T's *credibility*, I assume, prevents $\overline{T}$ from moving toward a final outcome in a game that is advantageous to $\overline{T}$ but disadvantageous to T. Should $\overline{T}$ contemplate such a move, T's threat will be to terminate the game at a (final) outcome disadvantageous to $\overline{T}$ (as well as himself).

I assume that T has the ability to carry out threats, which is his T-power. This is the power to hurt,[4] which T may threaten to use, but will actually use, only if his threat is ignored. In sum, while T's continuing involvement in a game is what motivates his use of threats, they are given force by his T-power, which enables T to terminate a game at a mutually disadvantageous outcome even though such termination is irrational for him (as well as $\overline{T}$) in any single play of the game.

To illustrate these concepts, again consider Chicken in Fig. 5.2. If neither player is engaged in repeated plays of this game, the game has a unique nonmyopic equilibrium, (3,3), which is also the common M-/S-power outcome. Now suppose that SB is T, so he can outlast P at any outcome disadvantageous to both, and both players know this. SB can move from (3,3) to (4,2) and then threaten P with *not* moving from (1,1) should P move there. This threat will induce P, given that he knows SB is engaged in repeated plays of the game and has T-power, to remain at (4,2). Because P knows that SB can endure the mutually worst outcome, (1,1), longer, P will not challenge SB's T-power by moving to (1,1). However, should SB lack credibility, P's move to (1,1) would ter-

---

4. Thomas C. Schelling, *Arms and Influence* (New Haven, CT: Yale University Press, 1966).

minate the game, providing an object lesson to P (as well as future—perhaps different—players) about the futility of challenging SB in Chicken.

By the symmetry of payoffs in Chicken, if P is the more powerful player (i.e., with T-power), he can obtain his best outcome (4) in this game. Whoever T is, by choosing his second strategy initially, he can either induce $\bar{T}$ to choose his first strategy, resulting in the best outcome for T and the next-worst for $\bar{T}$, if he is credible; or, if T is not credible, and $\bar{T}$ also chooses his second strategy initially, T can, by holding out at (1,1), force $\bar{T}$ to move first, which has the same effect as $\bar{T}$'s choosing his first strategy initially. In either event, the nonmyopic equilibrium, (3,3), is rendered vulnerable by one player's engagement in repeated play and his invocation of threats that, if necessary, he can and will carry out. Time, as Chicken demonstrates, may not heal all wounds; instead, it may exacerbate them.

Clearly, repeated play of Chicken puts T in the "driver's seat," committed to not swerving (i.e., choosing his second strategy) if, in the usual interpretation of this game, he is on a collision course with $\bar{T}$, who also chooses his second strategy.[5] This reckless behavior is deterred, however, if T's threat is recognized as credible, forcing $\bar{T}$ to "chicken out."

Curiously, the possession of threat power does not always help one player beat his opponent but, rather, may help both to extricate themselves from an unfortunate situation. Again consider Prisoners' Dilemma, shown in Fig. 5.2. Played once, this game has two nonmyopic equilibria, (2,2) and (3,3)—and is in fact the only 2 × 2 ordinal game to have more than one—but (3,3) is the common M-/S-power outcome, as in Chicken.

Repeated play of Prisoners' Dilemma undermines the stability of (2,2) at the same time that it confers on (3,3) the status of the "threat outcome." Assume, as before, that SB is T and the initial outcome is (3,3). If SB announces to P, "should you move to (1,4), I will move to (2,2) and not switch from my second strategy," this threat will be credible: P's two worst outcomes, one of which [presumably (2,2)] he would be stuck with if he departed from (3,3), are associated with SB's second strategy. Moreover, not only would P prefer (3,3), but so would SB.

Like P, SB (as T) would have no incentive to move to (4,1),

---

5. Steven J. Brams, *Game Theory and Politics* (New York: Free Press, 1975), pp. 39–40; and Steven J. Brams, *Paradoxes in Politics: An Introduction to the Nonobvious in Political Science* (New York: Free Press, 1976), pp. 114–115.

because this would simply induce P to move to (2,2), from which SB would not depart. Furthermore, SB can prevent P from choosing his second strategy initially by saying, "if you choose this strategy, I'll move to, and stay at, my own second strategy." Thereby both players would be motivated to choose their first strategies initially and not depart from (3,3).

Although one player's T-power singles out the Pareto-superior nonmyopic equilibrium as *the* outcome that would be chosen by rational players in Prisoners' Dilemma, Chicken illustrates how a player's T-power can undermine the unique (Pareto-superior) nonmyopic equilibrium. Hence, this kind of equilibrium, even when it is not Pareto-inferior—as is (2,2) in Prisoners' Dilemma—may be vulnerable to T-power.

Manifestly, T's ability to use threats effectively—that is, to ensure a better outcome than is possible without them—depends on the preferences of *both* players and their ability to implement desirable outcomes, according to rules I–IV (Section 4.4), in any single play. To explore this interdependence more systematically, consider the 2 × 2 ordinal outcome matrix in Fig. 6.1, in which each player is assumed to be able strictly to rank the four outcomes from best to worst. (The subsequent rather abstract analysis will be illustrated shortly.)

Assume that T desires to implement an outcome in row i and column j, (a_{ij}, b_{ij}), for some $i = 1$ or 2 and some $j = 1$ or 2, as the final outcome of the game, but he thinks that $\overline{T}$ might move from (a_{ij}, b_{ij}). Further suppose that to deter $\overline{T}$ from making this move, T threatens to force the termination of the game at some outcome $(a_{mn}, b_{mn}) \neq (a_{ij}, b_{ij})$. Call T's threat *real* (against $\overline{T}$) iff (if and only if), when carried out, it worsens the outcome for $\overline{T}$. T's threat is *rational* (for himself) iff, when successful in deterring $\overline{T}$, it improves T's outcome. A threat which is both real and rational is *credible* if it satisfies certain additional conditions.

To determine the conditions under which T has credible

Figure 6.1   *Ordinal 2 × 2 Outcome Matrix*

	P ($\overline{T}$)	
SB (T)	(a_{11}, b_{11})	(a_{12}, b_{12})
	(a_{21}, b_{21})	(a_{22}, b_{22})

Key: The outcome in row i and column j is denoted (a_{ij}, b_{ij}), where $i, j = 1, 2$.

threats, suppose, without loss of generality, that SB is T. Further suppose that SB threatens termination of the game at (a_{mn}, b_{mn}) in order to prevent P from moving from (a_{ij}, b_{ij}). Clearly, this implies that $b_{ij} < 4$—that is, (a_{ij}, b_{ij}) is not P's best outcome, for if it were, P would have no incentive to depart, making SB's threat superfluous. Now SB's threat is real iff

$$b_{mn} < b_{ij}; \tag{6.1}$$

it is rational iff

$$a_{mn} < a_{ij}. \tag{6.2}$$

Combining conditions (6.1) and (6.2), a necessary (but not sufficient) condition for SB's threat to be credible is that there exist a Pareto-inferior outcome:

$$(a_{mn}, b_{mn}) < (a_{ij}, b_{ij}), \tag{6.3}$$

which is taken to mean that the outcome on the left-hand side of the inequality is worse for both players than that on the right-hand side.

If T has a credible threat, $\overline{T}$ would not move from (a_{ij}, b_{ij}) only to accept Pareto-inferior outcome (a_{mn}, b_{mn}) as the final outcome. For this reason, I call (a_{mn}, b_{mn}) T's *breakdown outcome*: it forces $\overline{T}$ to comply with T's threat to avoid such an outcome, from which T will not depart. Instead, $\overline{T}$ is encouraged to stay at Pareto-superior outcome (a_{ij}, b_{ij}), which I call T's *threat outcome*; call T's strategy associated with this outcome his *threat strategy*. In contrast, call T's strategy associated with his breakdown outcome his *breakdown strategy*.

Because T would not move away from breakdown outcome (a_{mn}, b_{mn})—even though it is worse for both players than threat outcome (a_{ij}, b_{ij})—it is necessarily the final outcome should $\overline{T}$ depart from (a_{ij}, b_{ij}). The cost to T of suffering a breakdown outcome, I assume, is the price he is willing to pay in any single play of a game to ensure his credible threat is not viewed as empty (i.e., a bluff) in future games. Henceforth, I assume that T's possession of a credible threat implies that it is *nonempty*—T will exercise it by moving to, or not moving from, his breakdown strategy.

Two qualitatively different kinds of credible threats can be distinguished, one requiring no move on the part of T (assumed to be the row player (SB in the subsequent analysis), and the other requiring that he switch strategies:

1. *SB would stay: threat and breakdown strategies coincide.* When $m = i$, SB's threat strategy and breakdown strategy are the

same. Hence, SB can threaten to stay at the Pareto-inferior outcome (a_{in},b_{in}) should P move there. Condition (6.3) becomes

$$(a_{in},b_{in}) < (a_{ij},b_{ij}), \qquad n \neq j, \tag{6. 4}$$

so the existence of an outcome Pareto-inferior to the other outcome in the same row is sufficient for SB to have a credible threat. Note also that SB can always implement his threat outcome by choosing his threat strategy initially.

Chicken in Fig. 5.2 illustrates this case. As I showed earlier, SB can implement his threat (and most-preferred) outcome, (4,2), by choosing his second strategy initially and threatening not to move from (1,1), his breakdown outcome, should P move there.

2. *SB would move: threat and breakdown strategies differ.* When $m \neq i$, SB's threat and breakdown strategies are different. In this case, SB can threaten to move to his breakdown strategy should P move from (a_{ij},b_{ij}). Since P can move subsequently, he can choose as the breakdown outcome, (a_{mn},b_{mn}), the better of his two outcomes associated with SB's breakdown strategy. Necessarily, then, $b_{mn} \geq 2$. But I showed earlier that $b_{ij} < 4$, and from condition (6.3), $b_{ij} > b_{mn}$. Taken together, these inequalities imply $b_{ij} = 3$ and $b_{mn} = 2$. Hence, b_{mk} $(k \neq n)$, P's ranking of the *other* outcome associated with SB's breakdown strategy, must be 1. In other words, P's two worst outcomes are associated with SB's breakdown strategy m.

Thus, in case (2), SB can threaten, should P move from his strategy associated with (a_{ij},b_{ij}), to force P to choose between his next-worst and worst outcomes by switching to his breakdown strategy m. As in case (1), SB can always implement his threat outcome by choosing his threat strategy initially; unlike case (1), however, the Pareto-inferior outcome that establishes the credibility of SB's threat is associated with SB's other (i.e., breakdown) strategy. In sum, whereas breakdown and threat strategies coincide in case (1), they do not in case (2), necessitating a move on the part of SB in case (2) to demonstrate that his credible threat is nonempty.[6]

Prisoners' Dilemma in Fig. 5.2 illustrates case (2). Because P's two worst outcomes are in row 2, this is a breakdown strategy with which SB can threaten P. Moreover, SB has reason to: not only does he prefer his better outcome in row 1, (3,3), to P's better outcome in row 2, (2,2), but he also prefers (3,3) to (1,4) in row 1,

---

6. While it is possible to provide formal conditions under which SB has a reason to threaten P, they are not very illuminating. Their significance is mostly algorithmic, and they can easily be deduced from the algorithm for determining threat outcomes given in note 8.

whereas the reverse is the case for P. Thus, (2,2) is SB's breakdown outcome, and (3,3) his threat outcome. As T, then, SB can implement (3,3) under threat that if P moves to (1,4), SB will move to (2,2) and not move from row 2, his breakdown strategy, as I informally argued earlier.

In Schelling's terms,[7] case (1) reflects a *compellent threat:* T compels $\overline{T}$ to accept the threat outcome by saying that he will not abandon his threat/breakdown strategy should $\overline{T}$ challenge his threat (i.e., depart from the threat outcome and move to the breakdown outcome). Case (2) reflects a *deterrent threat:* T deters $\overline{T}$ from moving from the threat outcome by saying that he will retaliate by switching to his breakdown strategy should $\overline{T}$ challenge his threat. The two types of threats are not mutually exclusive and, indeed, may yield different threat outcomes in the same game. Presently I shall show that whenever the two threat outcomes differ, T will prefer the one associated with his deterrent threat.

The previous analysis allows one to state some propositions about threat strategies and outcomes in 2 × 2 ordinal games. The first follows immediately from condition (6.3):

i. *If a game contains only Pareto-superior outcomes, neither player has any threat strategies. In particular, neither player can credibly threaten his opponent in a total-conflict (constant-sum) game.*

Thus, the Pareto-inferiority of at least one outcome is necessary for the effective exercise of threat power in a game. Provided that the sufficient conditions given under cases (1) and (2) are met, the Pareto-inferior outcomes become breakdown outcomes for T. Furthermore, as indicated for each case:

ii. *T can always implement his threat outcome.*

Recall that by initially choosing his threat strategy, T can force $\overline{T}$ either to choose between T's threat and breakdown outcomes [case (1)]; or to risk T's choosing his breakdown strategy, containing $\overline{T}$'s two worst outcomes, unless $\overline{T}$ accedes to T's threat outcome [case (2)]. In either event, T's threat outcome is Pareto-superior to his breakdown outcome, so $\overline{T}$ will, if rational, choose his strategy associated with his threat outcome at the start.

Finally, it is not difficult to establish:

iii. *If T's compellent and deterrent threat outcomes differ, he will always prefer his deterrent threat outcome.*

7. Schelling, *Arms and Influence.*

To see this, note that if deterrent and compellent threat outcomes are different, they must appear in different rows (columns) of the outcome matrix. It immediately follows that the compellent threat outcome must be the breakdown outcome of the deterrent threat; consequently, condition (6.3) requires that it must be Pareto-inferior to the deterrent threat outcome.

There is exactly one game, shown in Fig. 6.2, in which a player has different (credible) compellent and deterrent threat outcomes. This player is SB, and his deterrent threat outcome, (4,3), is Pareto-superior to his compellent threat outcome, (3,2). On the other hand, (4,2) is SB's compellent threat outcome in Chicken in Fig. 5.2; (3,3) is not, however, a deterrent threat outcome for SB because the breakdown outcome, (4,2), is better, not worse, for him.

Two additional propositions that extend the previous analysis to $m \times n$ games larger than 2×2 are worth mentioning. They both give sufficient conditions for the implementation of the threatener's *best outcome:*

iv. *If T has "junk" strategy, all of whose outcomes are Pareto-inferior to his best outcome, associated with any other strategy, then he can induce the choice of this best (Pareto-superior) outcome with a deterrent threat.*

Clearly, T would prefer his best outcome to any outcome associated with the junk strategy. Moreover, T can ensure this outcome under threat of switching to his junk strategy.

"Brinkmanship," as commonly interpreted in superpower confrontations, might be thought of as a junk strategy. It is extremely dangerous, of course, to threaten destruction of the world with nuclear weapons, but it would not necessarily be irra-

Figure 6.2 *Only 2 × 2 Ordinal Game in which SB Has Both a Compellent and a Deterrent Threat*

	P	
SB	(2,4)	(4,3)^d
	(1,1)	(3,2)^c

Key: $(x,y) = (SB,P)$
4 = best; 3 = next best; 2 = next worst; 1 = worst
c = compellent threat outcome of SB
d = deterrent threat outcome of SB

tional if T thought he could survive and thence maintain supremacy over T̄ in future confrontations (if there were any!).

A compellent threat may also be used to cow an opponent in an $m \times n$ game:

v. *If T has a "carrot-and-stick" strategy, one of whose outcomes is best for him and Pareto-superior to all other outcomes associated with this strategy, he can induce this outcome with a compellent threat.*

A carrot-and-stick strategy differs from a junk strategy in having exactly one Pareto-superior outcome. If this outcome is best for T, his threat will induce its choice. An example of such a strategy might be one in which T threatens to punish T̄ severely (at a cost to T as well) unless T̄ makes certain concessions.

Note, however, that with a junk strategy, T can ensure his best outcome *whatever* other strategy it is associated with (i.e., wherever it lies in the outcome matrix). With a carrot-and-stick strategy, by contrast, T's best outcome must be the unique Pareto-superior outcome (the "carrot") associated with this strategy in order to guarantee its choice.

Thus, the possession of a junk strategy is, in a sense, a more powerful weapon than a carrot-and-stick strategy, because T can use it to implement his best outcome, wherever it lies in the outcome matrix, with a deterrent threat. By comparison, T's best outcome *must* be the Pareto-superior outcome in the carrot-and-stick strategy in order for him to implement it with a compellent threat.

To return to the 78 2×2 strict ordinal games, an algorithm can be used to determine which have threat outcomes and what they are.[8] Setting aside the 21 no-conflict games that contain a mutually best outcome, the remaining 57 2×2 games can be classified according to whether neither player has a threat strategy (11 games), one does (28 games), or both players do (18 games). Prisoners' Dilemma and Chicken fall into the last category, as I have shown, with the cooperative (3,3) nonmyopic equilibrium in the

---

8. If a 2×2 ordinal game has a threat outcome(s)—compellent or deterrent—it can be determined for any particular game as follows:
(1) Find i and j such that $a_{ij} = 4$.
(2) If $b_{ij} = 4$, stop: no threats are necessary to obtain a_{ij}.
(3) If $b_{ij} = 1$, go to (8): no threats can induce a_{ij}.
(4) Find (a_{mn}, b_{mn}) that satisfies condition (6.3). If none exists, go to (7).
(5) If $m = i$, (a_{ij}, b_{ij}) is SB's compellent threat outcome.
(6) If $m \neq i$, and $b_{mn} = 2$, stop: (a_{ij}, b_{ij}) is SB's deterrent threat outcome.
(7) If $a_{ij} = 3$, stop: SB has no threat outcomes.
(8) Find i and j such that $a_{ij} = 3$. Go to (2).

former game sustained by either player's T-power. But in Chicken, the (3,3) nonmyopic equilibrium is undermined by T-power, with SB (as the row player) able to implement (4,2) if he is T. Thus, T-power is more potent than M- or S-power in this game, wherein (3,3) is both the M-power and S-power outcome.

6.4. Threat Power in the Punishment Game: When Immortality Is Decisive in an Asymmetrical Game

Both Chicken and Prisoners' Dilemma are symmetrical games, so what is true for one player is also true for the other. In particular, because the goals of the players in these games are the same, their preferences are mirror images of each other, even if one player (presumably SB) has any of the attributes so far discussed: omniscience, omnipotence, or immortality.

Normally, though, it seems reasonable to suppose that there is an asymmetry in preferences between SB and P. However, in none of the theological games with asymmetrical preferences discussed so far—the Revelation Game, the Knowability Games (Prisoners' Dilemma with symmetrical preferences excepted), the Testing Game, or the Commitment Game—can SB, with T-power, improve upon the outcome he can implement with either M-power or S-power. In fact, in the Revelation Game (Fig. 5.1), SB has no threat outcome and thus could not assuredly avoid the Pareto-inferior (2,3) outcome with T-power, much less implement (4,2) or even (3,4).

But there are asymmetrical games in which SB's T-power is uniquely effective (in a nonperverse way): neither M-power nor S-power in these games makes a difference in favor of the player who possesses it. Indeed, in the game I shall describe, which I call the Punishment Game, only T-power enables SB to induce his best outcome, whereas the other kinds of power lead to his next-best or next-worst outcome.

In the Punishment Game I assume P has two strategies—sin or don't sin—and SB has two strategies—punish or don't punish P. The goals of the players are assumed to be the following:

P: (1) Primary goal—wants not to be punished
(2) Secondary goal—prefers to sin

SB: (1) Primary goal—wants to punish/not punish according to whether P sins/doesn't sin

(2) Secondary goal—prefers punishment over
    nonpunishment

The goals I have assumed of P—principally to avoid punishment, secondarily to sin—seem to me relatively uncontroversial. Implicit in this ordering for P is that sinning is pleasurable, particularly if he can escape punishment, his top priority. His overriding interest in avoidance of punishment is reflected in his two best outcomes being associated with nonpunishment by SB, his two worst with punishment. Sinning only establishes an ordering between these two best and two worst outcomes associated with each of SB's strategies.

SB, on the other hand, wants first to see justice done by following a tit-for-tat policy. More controversial is his secondary goal of preferring punishment. The reason, I think, for this preference on the part of the Old Testament God is rooted in His immortality:

> He wants to establish His credibility quickly. This is not so important in the immediate context as in the future games He will play or influence.
>
> Right from the start, therefore, God punishes all His adversaries—Adam, Eve, and the serpent—with a vehemence. The murder of Abel, which God instigates, follows soon thereafter. The stigma that Cain is forced to bear for this sin is designed not just to teach him a lesson but also to serve as a deterrent to all the people whom the peripatetic Cain encounters, or who later hear about the punishment.[9]

The outcome matrix, based on the postulated goals, is shown in Fig. 6.3. The T-power outcome for SB—which could be considered the result of a third kind of omnipotence, but which I prefer to link to immortality for reasons given in Section 6.2—is (2,4). Note that, unlike previous games, SB is the column player and P is the row player in Fig. 6.3, which is a switch that facilitates an expansion of SB's strategies that I shall describe shortly.

SB's threat outcome in the Punishment Game is a compellent one: by choosing punish, he can force P to accede to the Pareto-superior (2,4) outcome in the first column. P likewise has a compellent threat outcome, (3,3), associated with his strategy of not sinning, but since I assume P is not the immortal player in this game, I shall not consider his T-power and its effects further.

It should be observed, though, that T-power is effective in the Punishment Game because each player can induce a better outcome for himself if he, rather than the other player, possesses it.

9. Brams, *Biblical Games*, pp. 175–176.

Figure 6.3 *Outcome Matrix of Punishment Game*

		SB	
		Punish	*Don't punish*
Sin		Punishment justified	No punishment unjustified	
		(2,4)ᔆᴮ	(4,1)
P	*Don't sin*	Punishment unjustified (1,2)	No punishment justified (3,3)ᴾ

$\leftarrow$ Dominant strategy

Key: $(x,y) = (P,SB)$

4 = best; 3 = next best; 2 = next worst; 1 = worst

SB = T-power outcome for SB

P = T-power outcome for P

Circled outcome Nash equilibrium

In contrast, the M-power outcome for both SB and P is (3,3), whereas S-power is effective but has the paradoxical effect, noted in Section 5.7 (the Punishment Game is the game in Fig. 5.7 with the players reversed), of hurting SB when he has S-power and helping him when P does.

Earlier I suggested that the sin/punishment outcome, (2,4), is common in the Bible. But neither SB's M- nor S-power can induce it; its implementation by SB requires T-power. Formally, then, it can be explained in terms of the unique advantage that T-power gives God/SB in the Punishment Game, which seems to offer a generic representation of God's retribution in the Bible—and maybe elsewhere.

But, it might be contended, (2,4) is the unique Nash (though not a nonmyopic) equilibrium in this game and the product of a dominant strategy choice by P. It would, therefore, be the outcome classical game theory would predict, though P might feel bitter that his dominant strategy induces his next-worst outcome and SB's best, while SB himself does not have a dominant strategy.

For this reason, P might try to commit himself to his (dominated) "don't sin" strategy to induce the choice of (3,3). However, without T-power (or M-power), his commitment (or persistence) would not prove sufficient to overcome the "paradox of inducement" in this game.[10]

---

10. For a description of this paradox and references to it, see note 13, Chap. 2, where, unlike here, SB is the player with the dominant strategy.

But P has a way out if the Punishment Game is played such that SB makes his strategy choice after P makes his (or SB is omniscient). This is not an unreasonable supposition if SB truly desires to tailor his punishment to P's sins, or lack thereof. Then the game is not one of simultaneous choices by the players but instead one in which P chooses first—to sin or not—and SB responds to this choice.

This sequence is depicted by the game tree in Fig. 6.4 and the revised 2 × 4 outcome matrix, in which SB's four strategies depend on P's prior choice. As shown in Fig. 6.4, while SB's two "regardless" strategies are unconditional choices that do not

Figure 6.4 *Game Tree and Outcome Matrix of Revised Punishment Game in Which SB Chooses Second*

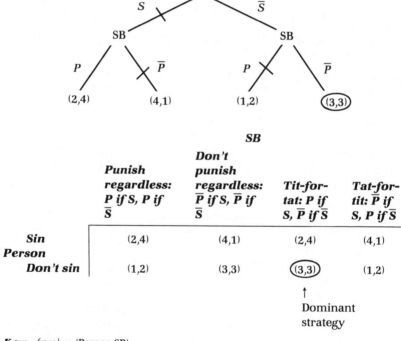

	SB			
	Punish regardless: P if S, P if $\overline{S}$	**Don't punish regardless: $\overline{P}$ if S, $\overline{P}$ if $\overline{S}$**	**Tit-for-tat: P if S, $\overline{P}$ if $\overline{S}$**	**Tat-for-tit: $\overline{P}$ if S, P if $\overline{S}$**
Sin	(2,4)	(4,1)	(2,4)	(4,1)
Person				
Don't sin	(1,2)	(3,3)	((3,3))	(1,2)

↑
Dominant
strategy

Key: (x,y) = (Person,SB)
   4 = best; 3 = next best; 2 = next worst; 1 = worst
   S = sin; $\overline{S}$ = Don't sin; P = Punish; $\overline{P}$ = Don't punish
   Nash equilibrium circled, which is also the outcome in the game tree that "survives" the slashes

depend on P's prior choice, "tit-for-tat" and "tat-for-tit" are conditioned on what P chooses first. SB's four strategies may be thought of as complete plans that specify his courses of action contingent on P's choices.

In this revised Punishment Game, SB has a dominant strategy (tit-for-tat); P does not but, anticipating SB's choice of tit-for-tat, he would choose "don't sin" since (3,3) is better for him than (2,4). Equivalently, backward induction on the Fig. 6.4 game tree demonstrates that (3,3) is the rational outcome because, as the slashes show, P (or Person, which I do not abbreviate to avoid confusion with "P" for SB's punish strategy) would choose $\bar{S}$ initially—to ensure (3,3)—rather than S, which would result in (2,4).

Recall that had P been able to await SB's choice in the Revelation Game, or had P simply been omniscient in this game, SB would have revealed himself and P would have believed in him, leading to the Pareto-superior (3,4) outcome in this game (see Sections 2.3 and 3.2). But SB, as I argued in Section 5.2, could do even better if he were M-powerful, obtaining (4,2), and also if he were S-powerful (see Section 5.6).

Similarly, if SB is T-powerful in the Punishment Game, he has no reason to choose second and induce only (3,3) in this game; instead, he can use his T-power to induce (2,4) in the Punishment Game. That is, by not deviating from his punish strategy, he can induce P to sin to salvage his best outcome from a bad situation.

As I indicated earlier, the biblical God was very much a threatening and punishing figure just after the creation of the world. This is not surprising: an immortal player who displays his threat power to the hilt can quickly establish his credibility and deter future antagonists. But, in some instances, inscrutability may serve SB's ends, and this I shall explore next.

6.5. Deception by the Superior Being

T-power, as I have demonstrated, is uniquely effective in inducing SB's best outcome in the Fig. 6.3 Punishment Game. But it is not effective in the Revelation Game (Fig. 5.1), in which only P—making the admittedly implausible assumption that he, not SB, has T-power—can induce his best outcome, (3,4), by threatening SB's two worst outcomes associated with P's "don't believe" strategy.

This is a deterrent threat; it might appear that SB has a compellent threat in his "reveal" strategy, but since (3,4) is P's best outcome, a threat by SB is not required to induce it. Moreover, (3,4) is only SB's next-best outcome—he would prefer to be able to induce (4,2).

Even to induce (3,4), SB would have to convince P that he would abandon his dominant "don't reveal" strategy in favor of revealing himself. Failing this, the Pareto-inferior Nash equilibrium, (2,3), would result, as noted in Section 2.3. But both M- and S-power, as I showed in Chapter 5, provide SB with an escape from this outcome, and not just to the Pareto-superior (3,4) outcome; M-power and S-power enable SB to implement (4,2), his best outcome in the Revelation Game.

Are there other means to this end that do not require either a steady alternation in player strategies until P gives up (M-power), or that SB holds off his strategy choices until P makes his choice and Rational Termination is operative (S-power)? There is in fact a third means for SB to achieve (4,2) in the Revelation Game, but it requires that P start with no information about SB's preferences— that SB appear inscrutable.

If SB's preferences are completely unknown to P, the Revelation Game would appear to P as it is depicted in Fig. 6.5, where *a*, *b*, *c*, and *d* are SB's unknown preferences. (I assume SB knows P's preferences, as shown.) In such a game of incomplete information, suppose that SB can falsely announce, or in some other

Figure 6.5 *Outcome Matrix of Revelation Game (Fig. 5.1) with Preferences of SB Unknown to P*

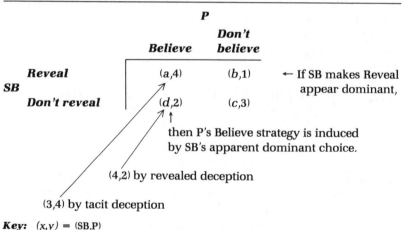

		P		
		Believe	**Don't believe**	
SB	**Reveal**	(a,4)	(b,1)	← If SB makes Reveal appear dominant,
	Don't reveal	(d,2)	(c,3)	

then P's Believe strategy is induced by SB's apparent dominant choice.

(4,2) by revealed deception

(3,4) by tacit deception

Key: (x,y) = (SB,P)

4 = best; 3 = next best; 2 = next worst; 1 = worst

way indicate without directly revealing himself, that he has a dominant strategy: he will reveal himself.

Because P has no knowledge of SB's preferences, I assume he will believe SB's announcement—he has no contrary information. Since P does not have a dominant strategy, he will choose his strategy that maximizes his own payoff in light of SB's (presumed) dominant strategy choice. Thus, P will believe, resulting in $(a,4)$.

By comparison, if SB had indicated that "don't reveal" was his dominant strategy, P would have chosen "don't believe," resulting in $(c,3)$. Not only is this outcome worse for P than $(a,4)$, but it is also worse for SB since, by assumption, $(a,4) = (3,4)$ and $(c,3) = (2,3)$ in the Revelation Game. Hence, SB will indicate his "reveal" strategy to be dominant, inducing P to believe and leading to $(a,4) = (3,4)$, where P does not know that $a = 3$.

This is a better outcome for SB, as well as P, than $(c,3) = (2,3)$, the Nash equilibrium. But it is not SB's best outcome, $(d,2) = (4,2)$.

Define $(a,4)$ to be SB's *tacit-deception* outcome. He can induce it by falsely announcing his first strategy to be dominant, when in fact his second is, in a game in which his preferences are unknown to P but he knows P's preferences, and P believes SB's (spurious) announcement.

SB's deception is tacit in this instance because P does not know that he has been deceived if SB in fact chooses his "reveal" strategy, consistent with his announcement. If, however, SB deviates from his announcement and actually chooses "don't reveal" in the play of the game, the result is $(d,2) = (4,2)$, SB's best outcome. This is the *revealed-deception* outcome because, since SB's announcement that his first strategy is dominant—inducing P to choose his first strategy—is inconsistent with his choice of his second strategy in the play of the game, he has exposed his deception to P. (It would obviously require a delicate maneuver on the part of SB to reveal his deception and not himself.) At the same time that he helps himself and hurts P, however, SB's behavior undermines the credibility of his future announcements, for he would probably not be believed again.

This is the price SB would pay in the Revelation Game in using revealed deception to ensure his best outcome (4). On the other hand, he can ensure his next-best outcome (3), as I showed, by adhering to his false announcement and so not revealing his (tacit) deception.

In general, revealed deception is a more powerful weapon than tacit deception, though the former kind of lying does not necessarily bring added benefit in some games.[11] For example, in the Commitment Game in Fig. 5.6, SB's announcement that his "await commitment" strategy is dominant—neither of his strategies in fact is—is sufficient to induce (4,3), his best outcome, without revealing his deception.

In some games, neither kind of deception helps. For example, in Prisoners' Dilemma (Fig. 5.2), the fact that P has a dominant strategy means that he cannot be swayed by SB into switching to his dominated strategy, for "don't compromise" is best whatever SB says or does. To be sure, in a game played according to sequential rules I–IV (Section 4.4), P can afford to take a chance and compromise in this game, knowing that if (4,1) is chosen and SB refuses to move to (3,3), he (P) can counter with a move to (2,2), which is worse for both players so presumably would not be necessary.

Implicit in the use of either tacit or revealed deception is that the game is a static one (i.e., only rule I applies): if SB can induce some better outcome by either kind of deception, the game will stay there. For if players could subsequently move at no cost, deception, in most cases, would be quite purposeless—there would be little point in SB's trying to induce a preferred outcome if P can depart from it immediately. Recall, however, that there are some games (e.g., the Commitment Game in Fig. 5.6) in which the initial outcome matters, for its choice precludes moves to certain other outcomes.

Depending on whether static of sequential rules or play are applicable in a particular situation, there are thus alternate routes to (4,2) for SB in the Revelation Game (and similar preferred outcomes in other games). The analysis here shows, in particular,

---

11. Of the 78 2 × 2 ordinal games, 17 are vulnerable to tacit deception and 27 to revealed deception; in 11 of the 17 vulnerable to tacit deception, including the Revelation Game, revealed deception leads to a better outcome than tacit deception. See Steven J. Brams, Deception in 2 × 2 games, *J. Peace Sci.* 2 (Spring 1977), 171–203; for an analysis of deception possibilities in other games, see Brams and Frank C. Zagare, Deception in simple voting games, *Social Sci. Res.* 6, 3 (September 1977), 257–272; Brams and Zagare, Double deception: two against one in three-person games, *Theory and Decision* 13, 1 (March 1981), 81–90. Applications of deception analysis to political games are given in Zagare, A game-theoretic analysis of the Vietnam negotiations: preferences and strategies 1968–1973, *J. Conflict Resolution* 21, 4 (December 1977), 663–684; Zagare, The Geneva Conference of 1954: a case of tacit deception, *Int. Studies Quarterly* 23, 3 (September 1979), 390–411; and Douglas Muzzio, *Watergate Games: Strategies, Choices, Outcomes* (New York: New York University Press, 1982), pp. 43–50.

that having a monopoly of information can be instrumental in SB's ability to deceive P into a false belief that leads to a better outcome for the deceiver, SB.

In my discussion of the Knowability Games in Section 2.4, I suggested means, other than deception, by which SB could become unknowable. One may question, however, how SB could deceive P with a false announcement in the Revelation Game and not reveal himself in the process of inveigling him.

My answer is that revelation may not be as all-or-nothing as previously implied. More specifically, an announcement, especially if its source is not clearly identified, hardly offers the kind of proof of existence of SB that some people may demand. For instance, despite the previously cited awesome and frightening events that occurred when Moses arrived with the Israelites at Mount Sinai (Section 5.4), his subsequent absence for forty days and nights on the mountain caused them to grow increasingly restive and finally idolatrous.

Thus, the signs of God's presence and power, manifest when they first reached Mount Sinai, proved insufficient to hold the people in thrall to the end. With the complicity of Moses's brother, Aaron, they revolted and made a golden calf. By that time, God's earlier displays of might and prowess had lost their force.

Revelation, in other words, may be a matter of degree. If God appears with sound and fury, as He did at Mount Sinai, He may likewise disappear like the morning fog as memories of Him fade. I think it is not impossible, then, to imagine SB's practicing deception by making false announcements at one time—or otherwise conveying misinformation without revealing himself—but then his presence dimming later.

Thereby the seeds of doubt are planted as SB retreats to non-revelation, though one might have felt earlier that he was a full-fledged player whose words, if not his deeds, had marked his unmistakable presence. But where is the trace now? This question, if no positive answer is apparent, can have devastating consequences on one's faith, particularly if one feels abjectly abandoned or trapped in some parlous state from which there is no escape. Then one's grief turns to anger and cynicism.

Next I undertake an analysis of optimal random strategies that may make SB appear incomprehensible, even though his preferences may be fully known. The game I shall use to illustrate these strategies is a generic one I call the Truth Game, in which truth, not SB, is the elusive quantity.

6.6. The Truth Game and the Problem of Evil

Before describing the Truth Game, it is useful to illustrate the optimality of random strategies in games of total conflict with a fifth Knowability Game. It will be remembered that one of the previous four Knowability Games (game II in Fig. 2.2), described in Section 2.4, was also one in which the best outcome for one player (4) was worst for the other (1), and the next-best for one (3) was next-worst for the other (2). But this game had a Nash (and nonmyopic) equilibrium that was the product of dominant strategies of both SB and P, so neither player had an incentive to deviate from his strategy associated with it.

In the fifth Knowability Game, shown in Fig. 6.6, I assume that the goals of the players are the following:

> **P:** (1) Primary goal—to confirm SB's knowability/ unknowability
> (2) Secondary goal—prefers that SB be knowable
> **SB:** (1) Primary goal—wants P's expectations of his knowability/unknowability disconfirmed
> (2) Secondary goal—prefers to be unknowable

As the arrows in Fig. 6.6 indicate, one player has an incentive to depart from every outcome, making this a thoroughly unstable game.

It is true that if SB had moving or staying power, he could implement (3,2), which is better for him and worse for P. But since this game contains no Pareto-inferior outcomes, SB does not have T-power, though his omniscience and P's awareness of it would also result in (3,2). In addition, SB could induce (2,3) through tacit deception and (4,1) through revealed deception, but unlike the M- and S-power outcomes, there is no reason for the players to stay at these initial outcomes or, for that matter, at the (3,2) omniscience outcome.

Assume that the ranks in Fig. 6.6 are actually numerical payoffs, or cardinal utilities, to the players. Then the Fundamental Theorem of Game Theory, or the famous Minimax Theorem, says that either player, by choosing each of his strategies at random with probability ½ in this game (e.g., by flipping a fair coin and letting heads signify "choose first strategy," tails "choose second

Figure 6.6   Outcome Matrices of Fifth Knowability Game and Truth Game

Fifth Knowability Game

		P		
		Investigate SB because expect knowability		**Don't investigate SB because expect unknowability**
SB	**Be knowable**	(1,4)	←	(3,2)
		↓		↑
	Be unknowable	(4,1)	→	(2,3)

Truth Game

		SB		
		Believe		**Don't believe**
P	**Tell truth**	(2,10)	←	(3,7)
		↓		↑
	Don't tell truth	(10,0)	→	(0,8)

Key: (x,y) = (row player, column player)
Arrows between outcomes indicate rational departures from each outcome by row (vertical) and column (horizontal) players

strategy"), can guarantee himself the *value* of 2.5, whatever strategy the other player chooses.[12]

The value, in other words, is what a player can ensure for himself by himself. Neither "pure" (i.e., single) strategy can guarantee this value—only a 50-50 "mixture" in this game can ensure, on an expected-value basis, this amount to each player, independent of what the other player does.

The Fundamental Theorem applies only to two-person constant-sum games (i.e., games of total conflict), in which the payoffs

12. John von Neumann and Oskar Morgenstern, *Theory of Games and Economic Behavior*, 3rd Ed. (Princeton, NJ: Princeton University Press, 1953). A relatively nontechnical explication of these concepts and underlying calculations for two-person constant-sum games is given in Brams, *Game Theory and Politics* (New York: Free Press, 1975), pp. 1–25.

to each player sum to the same constant at each outcome (5 in the case of the fifth Knowability Game). This severely limits the applicability of the theory in the analysis of superior beings, for almost all the games I have discussed are games of partial conflict.

The idea of using mixed or randomized strategies, nevertheless, is an important one. If SB and P are totally at odds with each other, as supposed in the fifth Knowability Game, and there is no equilibrium outcome ("saddlepoint" in a constant-sum game), then it may make sense, say, for SB not to make his strategy choices in any predictable manner. (If he did, this information could be exploited, as I shall show explicitly in the Truth Game.)

Not only does an optimal mixed strategy minimize SB's maximum loss, which is why the Fundamental Theorem is also called the Minimax Theorem, but it does so without his having to invoke any of his putative superior powers. SB may, for any number of reasons, not want to assert his control in any obvious way; the randomness of a mixed strategy allows him to be inconspicuous if not incomprehensible.

In mixing his choices in the fifth Knowability Game, sometimes appearing knowable and sometimes not, SB would undoubtedly present a very confusing picture to P. P could as well expect or not expect knowability, or mix his own choices—his payoff would be the same, on the average, if SB played his optimal mixed strategy. This is precisely when the inconsequentiality of P's choice is most apparent, for it does not have any material effect on his payoff if SB himself is acting arbitrarily, albeit in an optimal way (the optimal probability mixture need not be 50-50, as I shall illustrate shortly).

The problem P faces in the fifth Knowability Game is reminiscent of his playing a one-person game against indifferent nature (see Section 2.5), though in a two-person constant-sum game, "nature" is in fact a player with its own payoffs. Consequently, in the fifth Knowability Game, SB has better and worse choices, so is not truly indifferent, but his best strategy—in the sense of guaranteeing a certain expected payoff—happens to entail random choices fixed by certain probability constraints (i.e., it is optimal for SB to choose each strategy with the same 0.50 probability).

The Truth Game in Fig. 6.6 is not constant-sum: both players can simultaneously do better [at (2,10)] or worse [at (0,8)] at different outcomes, so what one player "wins" the other player does not necessarily "lose." The particular payoffs assumed were chosen to illustrate certain points that follow from the calculations to be described shortly; note that the best and worst payoffs of each player are 10 and 0, respectively. Like other games analyzed in this book, the underlying player rankings of the four outcomes can be specified by primary and secondary goals, to be given below.

In the Truth Game I assume that P must choose between telling the truth and lying, and, having made this choice, SB must then decide to believe him or not. Although I assume that P chooses prior to SB, this game cannot be modeled, like the revised Punishment Game in Fig. 6.4, as one of sequential play with perfect information. This is because I assume that SB cannot tell for certain whether P was truthful or not and thus cannot respond to P's strategy by always choosing his own better outcome in the first or second row. Since SB is unsure which strategy P chose, this is technically a game of *imperfect information*, though I assume that both players have complete information about the Fig. 6.6 payoffs.

Implicit in these payoffs are the following goals:

P: (1) Primary goal—wants to *hide* the truth, or lack thereof (two best outcomes off the diagonal)
(2) Secondary goal—prefers to be believed

SB: (1) Primary goal—wants to *discover* the truth, or lack thereof (two best outcomes on the diagonal)
(2) Secondary goal—prefers P to be truthful

Clearly this is a hide/discover-the-truth game, with a secondary emphasis on P's desire to be believed and SB's desire that P be truthful. In Section 4.1, I indicated instances of this kind of game being played in the Bible, where, for example, God asks questions of Adam and Eve and Cain to try, at least ostensibly, to ferret out the truth about their sins.

Like the fifth Knowability Game, there is no stability in the Truth Game, as the arrows indicating cyclical preferences over the four outcomes in Fig. 6.6 make evident. Furthermore, SB does not have a compellent or deterrent threat in the Truth Game, but he can implement his best outcome, (2,10), with M- or S-power.

Yet, it is hard to conceive what M- or S-power would mean in the Truth Game, wherein the question of discovering the truth seems to hinge less on SB's power and more on his detection capability. To make this latter idea more precise, assume SB has partial omniscience and can detect P's strategy choice correctly, whatever it is, with probability p (a related calculation for Prisoners' Dilemma was given in Section 3.6).

If P knows p, too, he can make an expected-payoff calculation (see Sections 2.5, 3.6, and 3.8) for being truthful (T) or not ($\overline{T}$), given that SB follows a tit-for-tat policy—believes P if he detects T, doesn't believe P if he detects $\overline{T}$. Thus, for P, the expected payoff, E, associated with each of his strategies is

$$E(T) = 2p + 3(1 - p) = 3 - p;$$

$$E(\overline{T}) = 10(1 - p) + 0p = 10 - 10p.$$

Lying is rational if $E(\overline{T}) > E(T)$, or

$$10 - 10p > 3 - p$$

$$p < \tfrac{7}{9} = 0.78.$$

For purposes of illustration, assume $p = \tfrac{3}{4} = 0.75$. Then it is rational for P to lie, yielding

$$E(\overline{T}) = (10)(\tfrac{1}{4}) + (0)(\tfrac{3}{4}) = \tfrac{10}{4} = 2.50.$$

Thus, when P lies, SB, using his detection equipment, will correctly detect $\overline{T}$ three-fourths of the time and incorrectly detect T one-fourth of the time; so his tit-for-tat policy will yield him

$$E(\text{tit-for-tat}) = (8)(\tfrac{3}{4}) + (0)(\tfrac{1}{4}) = \tfrac{24}{4} = 6.00.$$

Hence, P does worse than his next-*best* payoff of 3, and SB does worse than his next-*worst* payoff of 7. Comparatively speaking, then, SB ranks his expected payoff lower than P does, but quantitatively SB does much better than P (assuming their utilities are measured on the same scale and can be compared).

Surprisingly, an extension of the mixed-strategy calculation to nonconstant-sum games like the Truth Game can help the hapless SB in this game.[13] Let

q = probability that SB chooses believe (B) if his detector indicates T;

r = probability that SB chooses B if his detector indicates $\overline{T}$.

Previously I assumed that $q = 1$ and $r = 0$ when SB followed tit-for-tat, but now I want to show that it is optimal in the Truth Game for SB to respond to his detector (still assumed to have reliability $p = \tfrac{3}{4}$, whatever P chooses) probabilistically rather than deterministically.

---

13. The calculations that follow were developed in collaboration with Morton D. Davis, for whose advice I am grateful. Although they are very different from the calculations developed in Vladimir A. Lefebvre, *Algebra of Conscience: A Comparative Analysis of Western and Soviet Ethical Systems* (Dordrecht, Holland: D. Reidel, 1982), Lefebvre also analyzes ethical structures that underlie good and evil. He shows how they differ fundamentally in Western and Soviet societies.

Let s = probability that P chooses T. Then for $0 \leq q,r,s \leq 1$, P's and SB's expected payoffs (after considerable simplification) are

$$E(P) = (\tfrac{1}{4})[s(12 - 13q - 31r) + 30r + 10q];$$

$$E(SB) = (\tfrac{1}{4})[s(-4 + 17q + 27r) - 24r - 8q + 32].$$

Now, by keeping $(12 - 13q - 31r)$ positive, SB can induce P to tell the truth, because this will raise $E(P)$, given $s > 0$.

In particular, if $r = 0$ and $q = \tfrac{12}{13} - \epsilon$, where ϵ is some arbitrarily small positive number, P should always tell the truth ($s = 1$), yielding

$$E(P) = (\tfrac{30}{13})^{+} = 2.31^{+}.$$

Similarly, SB's expected payoff will be

$$E(SB) = (\tfrac{118}{13})^{-} = 9.08^{-}.$$

The superscripted "$+$" and "$-$" signs indicate that these payoffs will be slightly more than and less than their numerical values, respectively, depending on the "ϵ-incentive" SB offers to P to tell the truth.

Observe that, keeping this incentive sufficiently small,

$$E(P) = 2.31^{+} < E(\overline{T}) = 2.50,$$

$$E(SB) = 9.08^{-} > E(\text{tit-for-tat}) = 6.00,$$

so SB does better, but P does not, when SB follows a more "sophisticated" policy than tit-for-tat. That is, when SB makes his strategy choices probabilistic—according to q and r given in the previous paragraph—he will do considerably better and, at the same time, marginally decrease P's expected payoff.

The key to this optimality calculation is that SB can induce P to choose T over $\overline{T}$ by deviating from a straight tit-for-tat policy. Unexpectedly, according to the calculated q and r optimality values given earlier, SB should only selectively follow the signals of his detector, even though it has reliability $p = \tfrac{3}{4}$. For example, if it indicates P chose $\overline{T}$, it is optimal for SB always to choose $\overline{B}$ (as expected), but if it indicates P chose T, it is optimal for SB to choose B only about $\tfrac{12}{13} = 92\%$ of the time.

Of course, when $p = 1$ (i.e., SB's omniscience is total), he should adhere punctiliously to tit-for-tat. But then P should always tell the truth, because his payoff will be 2 versus 0 for lying; this, in turn, gives SB his highest payoff of 10.

The final surprise in these calculations is that when $p = \tfrac{3}{4}$, SB, treating the Truth Game as constant-sum in *his* payoffs, can guarantee himself an expected payoff of $^{128}/_{17}$ = 7.53, whatever P does; in this case, P does best to choose T, giving him an expected payoff of $^{48}/_{17}$ = 2.82. Note that these expected payoffs are greater for *both* players than their expected payoffs when SB follows tit-for-tat deterministically.

In sum, SB should never follow tit-for-tat blindly in the Truth Game (unless his omniscience is total). He should instead follow a probabilistic choice rule, whether to guarantee himself a certain value independent of what P does, or to try to do even better by inducing P to tell the truth.

These results extend the idea of a mixed-strategy solution in two-person constant-sum games like the fifth Knowability Game to two-person nonconstant-sum games like the Truth Game, in which SB has some detection capability. The fact that it is sometimes optimal for SB to ignore the signal he receives suggests that optimal randomized strategies in games may be more pervasive than has heretofore been thought. This is true not only of games involving SB but also of secular games between a potential liar and a detector, which perhaps gives the Truth Game generic qualities that most of the other games discussed in this book do not possess.

The optimality of randomized strategies has particular poignancy in games involving SB. Insofar as SB has partial omniscience and thus can assume the role of a detector, the results for the Truth Game show that it is not necessarily advantageous for him to reward and punish precisely according to what he detects. Instead, at least occasionally, he should believe in those perceived to be untruthful, punish the righteous, and in general appear to act callously and ignominiously.

In the Bible, there are many instances of a warped kind of justice. For example, a favored character like David for the most part gets away with adultery and even murder, whereas those out of favor, like his predecessor, Saul, are treated very shabbily. "Simple clarity and directness are not God's chosen means."[14]

Is this fair? I think not. In the contemporary world, too, the innocent often suffer as the guilty go unpunished. Justice, to put it charitably, is not perfect.

Are corruption, lying, and other seamy and wanton activities in the world testimony to the *non*existence of a benevolent God? One lesson of the Truth Game, I think, is that if there is an overarching intelligence with partial omniscience, it may be rational for him—even when he adopts the worthy goal of discovering the

---

14. Frederick Sontag, *The God of Evil: An Argument from the Existence of the Devil* (New York: Harper & Row, 1970), p. 134.

truth—sometimes to ignore what he detects. In other words, his behavior should, on occasion, be deliberately arbitrary.

This finding needs much more refinement, detailed analysis, and systematic development than I can give it here. I wish to stress, though, that I am not condoning evil, as evinced in arbitrary behavior. Rather, I am trying to explain why it may be rational on occasion and why "alongside justice [and] . . . righteousness there is wickedness" (Eccles 3:16).

Disturbing as this kind of behavior is, one's normative position does not extirpate it. All too plainly, evil and suffering exist in the world. What I have tried to do in this section is sketch, with some suggestive calculations, an alternative explanation for arbitrary behavior, given a SB and a P with certain utilities in a game in which SB has partial omniscience and seemingly exemplary primary and secondary goals.

The fact that such behavior may indeed be purposeful, in a strange sort of way, is perhaps upsetting if not morally repugnant. One way out is to say that, because we have free will and God is therefore not all-powerful, He cannot prevent our behaving unconscionably toward one another.

This is the tack that Kushner takes in his book, *When Bad Things Happen to Good People*.[15] I do not find it edifying, despite Kushner's valid criticisms of traditional theological positions. The problem with this kind of rational explanation of evil, in my opinion, is that even endowing SB with unimpeachable goals and limited powers, which largely preserve P's free will, it still may be rational for SB to be arbitrary.[16] Assuming there are no limits on God's powers, Lord Byron raised the plaintive question in *Cain* (1821): "Because He is all-powerful, must all-good follow?"

One does not understand the problem of evil by trying to sweep it under the rug. First of all, it is too glaring to ignore—capriciousness seems to be everywhere. But more to the point, I think we must pause to consider that if arbitrariness has a rational basis, incomprehensibility may have its place in a higher design.

---

15. Harold S. Kushner, *When Bad Things Happen to Good People* (New York: Schocken, 1981).

16. In *When Bad Things Happen to Good People*, Kushner quotes Job (p. 41) to the effect that there are "no rules" in understanding God: "He snatches away—who can stop Him? Who can say to Him, 'What are You doing?'" (Job 9:12). Yet, as I showed, arbitrary and seemingly unfathomable behavior is entirely consistent with rules of those games that prescribe random strategy choices. That God in fact makes these choices, perhaps for our own good, I cannot say. However, arbitrariness itself is certainly not inexplicable behavior in games; indeed, it may be optimal to use subterfuge.

However, I know of no way to determine whether it stems from an indifferent and desultory nature, or a concerned and calculating SB optimally randomizing his choices.

Unfortunately, neither explanation provides much solace to a P who tries to reconcile the noble and the ignoble in the world. Perhaps the most that can be said for the first is its consistency with the "banality of evil" view; the second is a more demonic view, though SB is not necessarily conceived to be evil incarnate but instead may have higher motives.

6.7. Conclusions

In this Chapter I have tried to go beyond the usual divine attributes of omniscience and omnipotence, which secular players may well have in watered-down form. Indeed, some readers may consider my game-theoretic definitions of these qualities hopelessly inadequate in capturing the profound differences that separate human from all-knowing and all-powerful beings.

There is nothing human about immortality, though, which is distinguishable in kind—not just degree—from mortality. But as with omniscience and M- and S-power, T-power in repeated games, and the willingness of SB to endure a Pareto-inferior outcome in the single play of a game, may not begin to express the unbridgeable gulf that separates living forever from dying.

If profane concepts and theoretical tools do not touch and illuminate transcendent qualities of superior beings, intellectual enterprises like this are doomed from the start. But to me the fact that T-power is effective in the Commitment Game when M-power and S-power are not indicates that there may be something peculiar about immortality in a game-theoretic setting, just as omniscience and omnipotence, as I have defined them, may also be uniquely effective. Similarly, tacit and revealed deception can help SB, though the latter kind of deception certainly could prove costly, especially because, as an immortal player, SB must concern himself with how future players view him.

If P is not in the dark about SB's preferences, as I assumed was the case in the deception analysis, it may be rational for SB to randomize his choices. The optimality of such mixed strategies is well known in two-person constant-sum games, like the fifth Knowability Game, but these games of total conflict are a special case. Of more general interest, I think, are games of partial conflict, like the Truth Game, in which SB wants to discover the truth and P wants to cover it up.

This is not a game unique to a SB with certain detection capabilities; it also characterizes secular conflicts in which the parties make choices that are partially predictable. Nevertheless, the finding that SB should randomize his choices in a particular way has ramifications for the bewildering problem of evil.

Specifically, the fact that the arbitrariness implicit in SB's randomized strategies can be explained in rational-choice terms may be abhorrent, especially if SB's goals seem beyond reproach. Not only does haphazard behavior make SB incomprehensible, but it also implies that he should not strictly follow a tit-for-tat retribution policy; instead, some chicanery may be called for.

This may result in wrongful punishment of the righteous, and undeserved rewards to the wicked, making SB appear unethical if not despicable. I conclude that though this behavior may be inconsistent with the supposed benevolence and rectitude of God, it is not necessarily irrational or misguided for certain goals one might attribute to an abstract superior being. The perhaps unanswerable question is how closely the latter resembles the former.

In pondering this question, it is worth recalling the biblical injunction:

> For as wisdom grows, vexation grows;
> To increase learning is to increase heartache.   (Eccles. 1:18)

I hope, in this chapter, I have not compounded heartache with a mathematical headache!

SEVEN

Superior Beings: They May Be Undecidable

7.1. Introduction

It is now time to summarize the copious results that game theory yields when superior beings confront ordinary beings in games (see Appendix). I wish to reiterate that these theoretical results are just that—conceptual guides for thinking about certain religious-theological-philosophical questions but not scientific findings supported by any kind of empirical evidence, even if couched in mathematical language. As I recapitulate the effects of SB's powers in games, I shall use them as a springboard to discuss what I identified as the "central question" in the first sentence of the Preface of this book: "If there existed a superior being who possessed the supernatural qualities of omniscience, omnipotence, immortality, and incomprehensibility, how would he/she act differently from us, and would these differences be knowable?"

This, in my opinion, is the most profound of existential questions one can ask. It leads us, via game theory, to a possible explanation of the problem of evil, based on the rationality of arbitrary behavior (given partial omniscience on the part of SB that is compatible with P's having free will), which I discussed in Chapter 6.

In this chapter, I will define and discuss "undecidability," which I shall tie to arbitrariness to argue that there are daunting problems we face in piercing the Great Mystery.

7.2. Undecidability

To begin the summary analysis, consider the effects of SB's *combined* superior qualities, based on his omnipotence and immortality. Unhappily, the picture would appear complex: in some games M-power is effective, in some S-power is, and in some T-power is. There is substantial but by no means complete overlap among these games. What complicates the picture further is that SB can do best with M-power in one game, S-power in another, and T-power in a third, so he might have to decide which kind of power to invoke when two or more of these superior qualities are effective.

Recall that effectiveness means that SB's M-power, S-power, or T-power induces a different outcome—and better in all but one case involving S-power—when he has it than when P does. When possession makes no difference, these qualities still may be useful to implement *some* outcome, such as the compromise (3,3) outcome in Prisoners' Dilemma or the Testing Game, which both kinds of omnipotence as well as immortality ensure. Both kinds of omnipotence also ensure (3,3) in Chicken, and M-power ensures (3,3) in the Punishment Game, but T-power upsets (3,3) in both games in favor of (4,2) for SB, so presumably SB would use it if he had all three kinds of power at his disposal.

Similarly, SB would use S-power over M-power in the Commitment Game, because it implements his best outcome, and M-power over S-power in the game in Fig. 5.7, because it leads to a better outcome for himself. Finally, he could use either kind of omnipotence but not T-power to implement his best outcome in the Revelation Game, which establishes that no attribute dominates the other two in all games.

To turn these positive results on SB's omnipotence and immortality around, consider games in which *none* of these superior qualities is effective. It so happens that this includes the bulk of the 78 2 × 2 ordinal games (52 in all), but in most of these SB can do quite well. For example, 21 of these games contain a mutually best (4,4) outcome, so it is reasonable to suppose that players, superior or not, would implement it.

Clearly, SB's superior qualities of moving, staying, and threat power are for nought when they are needed but not helpful. Two criteria of need that would seem hard to dispute—though there certainly are others—are: (i) SB does worse than P, and (ii) the best outcome SB can ensure is his next-worst (2). In fact, all games that satisfy criterion (ii) satisfy (i) as well, but not vice versa, so (ii) is really the key requirement.

What does it mean? Despite SB's superior qualities, he cannot ensure a better outcome for himself than 2, the security level of an ordinary player. Since any player can implement this outcome simply by choosing his strategy that is *not* associated with his worst outcome (1), SB gains absolutely nothing from having M-, S-, or T-power. It happens to be true that P does comparatively better than SB in all games that satisfy criterion (ii)—if interpersonal comparisons of player ranks (or utilities) are allowed. But such comparisons aside, the important point is that no kind of power permits SB to escape his next-worst outcome.

I call games in which the three kinds of power cannot prevent SB's next-worst outcome *undecidable*. The reason for using this terminology is that if SB were a player in such a game, an observer could not decide from the outcome implemented whether he was superior; the outcome would be the same if SB were an ordinary player conservatively choosing his strategy not associated with his worst outcome.

This is not the same meaning that "undecidability" has in logic, wherein it signifies that neither a statement nor its negation can be deduced from the axioms of a consistent mathematical system—that is, both are nontheorems within the system.[1] Its import, nevertheless, is the same: from the outcome of the game alone, one

---

1. A mathematical system is consistent if no two theorems can be derived that contradict each other. For further details, see Ernest Nagel and James R. Newman, *Gödel's Proof* (New York: New York University Press, 1968); Mark Kac and Stanislaw M. Ulam, *Mathematics and Logic: Retrospect and Prospects* (New York: New American Library, 1969); Edna E. Kramer, *The Nature and Growth of Modern Mathematics*, Vol. 2 (Greenwich, CT: Fawcett, 1970); Douglas R. Hofstadter, *Gödel, Escher, Bach: An Eternal Golden Braid* (New York: Basic, 1969); and Rudy Rucker, *Infinity and the Mind: The Science and Philosophy of the Infinite* (Boston: Birkhäuser, 1982). For arguments that Gödel's Theorem is not just a "logician's trick," see Gina Kolata, Does Gödel's Theorem matter to mathematics? *Science* 218, 4574 (19 November 1982), 779–780. The connection between Gödel's Theorem and theology is made in Howard Eves, *Great Moments in Mathematics (after 1650)* (Washington, DC: Mathematical Association of America, 1981), Lecture 38 (Mathematics as a branch of theology, pp. 200–208); Eves quotes from Frank DeSua, Consistency and completeness—a resumé, *Am. Math. Monthly* 63, 5 (May 1956), 295–300, to the effect

cannot determine whether one player is M-powerful, S-powerful, or T-powerful, because he could not have improved the outcome if he were. These outcomes are like nontheorems in that they cannot be attributed to SB's superior abilities, which are insufficient to change the outcome to something SB prefers, just as the axioms of a system are insufficient to establish an undecidable statement or its negation as a theorem.

Certainly, SB would have reason to try to improve the outcome if it were his next-worst. On the other hand, he would have no reason to try if he could induce his best outcome without using any superior qualities, which shows the significance of criterion (ii) in this test: a game is undecidable not only because the three kinds of power make no difference in the outcome SB can induce, but also because SB desires a *better* outcome.

Now SB can always avoid his worst outcome (1) by moving from it, as allowed by the sequential rules of play. The question of when SB can or cannot improve upon his next-worst (2) and next-best (3) outcomes I shall now take up.

A remarkably simple *undecidability rule* characterizes all 2 × 2 ordinal games in which SB is stuck with his next-worst outcome: SB's two worst outcomes (1 and 2) are associated with one strategy of P, and this strategy includes P's two best outcomes (3 and 4). One such game is illustrated in Fig. 7.1, with SB's two worst outcomes, and P's two best, associated with P's first strategy. Necessarily, P's two worst outcomes, and SB's two best, are associated with P's second strategy.

Naturally, P will choose his dominant first strategy in this game, and SB will choose his first strategy in order to obtain his

---

that only mathematics among disciplines carries within it—via Gödel's Theorem—a rigorous demonstration that its foundations rest on a kind of religious faith.

For arguments in the political science literature that there is an undecidability problem in ascertaining the exercise of power, see *Political Power: A Reader in Theory and Research*, ed. Roderick Bell, David V. Edwards, and R. Harrison Wagner (New York: Free Press, 1969), particularly the two articles by Peter Bachrach and Morton S. Baratz, Two faces of power (pp. 94–99) and Decisions and nondecisions: an analytical framework (pp. 100–109). These articles were originally published in, respectively, *Am. Political Sci. Rev.* 56, 4 (December 1962), 947–952; and *Am. Political Sci. Rev.* 57, 3 (September 1963), 632–642. After I completed this book, an article by J. P. Jones, Some undecidable determined games, *Int. J. Game Theory* 11, 2 (1982), 63–70, appeared. Jones proves that there are rather simple two-person win-lose games with perfect information in which one player has a winning strategy, but it cannot be decided which one; this kind of undecidability is different from determining whether one is playing against a superior being but, in my opinion, equally fascinating and perplexing.

Figure 7.1   *Outcome Matrix of Undecidable Games*

	P	
	(2,3)	(4,1)
SB	$\updownarrow$	$\updownarrow\updownarrow$
	(1,4)	(3,2)

Key: (x,y) = (SB,P)

4 = best; 3 = next best; 2 = next worst; 1 = worst

Arrows indicate three possible pairwise interchanges of preferences for the two players, once *SB*'s first-column preferences are fixed

next-worst and avoid his worst outcome. It is easy to show that neither kind of omnipotence, nor immortality, enables SB to induce an outcome for himself better than (2,3) in this game.

The *class* of all undecidable games is indicated by the arrows showing three possible pairwise interchanges of the two sets of 3 and 4, and 1 and 2, in the Fig. 7.1 game. Once the 2-1 ordering for SB is fixed in the first column, the three interchanges demonstrate that there are 2 × 2 × 2 = 8 distinct 2 × 2 games that satisfy the undecidability rule. The second Knowability Game in Fig. 2.2 is one such game, though it is P in this game who is "helpless" according to criterion (ii): if he and SB reversed roles in this game, SB's superior qualities would not extricate him from (3,2).

This example illustrates the fact that the eight undecidable games indicated in Fig. 7.1 are undecidable only when SB is the row player. If he is the column player, with or without superior qualities he can implement his next-best or best outcome, violating criterion (ii).

All eight undecidability games, therefore, are asymmetric: unlike the symmetric games of Chicken and Prisoners' Dilemma, the goals/preferences of the two players are different, and consequently so is the strategic situation each player (SB or P) faces. The undecidable (2,3) or (2,4) outcome in the upper left-hand corner of each game in Fig. 7.1 is a Nash equilibrium in all eight games, but it is a nonmyopic equilibrium in only four.

Two of the eight games are games of total conflict, and both have a Nash equilibrium (saddlepoint). There is a structurally distinct 2 × 2 ordinal game of total conflict without a Nash equilibrium, illustrated by the fifth Knowability Game in Fig. 6.6, that is not undecidable. The other six games, though not total-conflict games, are partial-conflict games that are "close" since every outcome is one of the top two for one player, the bottom two for the other.

One might suspect, at first glance, that games like the unde-cidability games, which reflect serious differences between the two players (they never rank any outcome the same), would be most vulnerable to the exercise of power. But the opposite is the case: SB is powerless to change the unsatisfactory outcome for himself in any of these games.

Omniscience and the ability to deceive—the other superior qualities discussed earlier—are also of no help. P will not depart from his dominant strategy if he is aware that SB can predict his choice, for it still is unconditionally best. Likewise with deception, if P does not know SB's preferences, again he will not be moved from his dominant strategy. Hence, neither what SB announces to be dominant for himself, nor P's awareness of his omniscience, can be exploited by SB in the undecidable games.

There are 16 games which meet criterion (i) but not (ii); in these, SB's superior qualities can ensure a (3,4) outcome, in which SB does next-best and P best. Although SB does not do as well as P, he is at least able to prevent the choice of his next-worst outcome.

As before, I would stress that SB obtains an inferior (here, next-best) outcome, not that he does worse than P. There are an additional seven games in which SB can ensure (3,3); since his ranking of this outcome is the same as P's, it presumably would not instill resentment in SB if interpersonal comparisons were allowed.[2] I discuss the resentment factor (which is not usually assumed to be incorporated in a player's payoffs) in Section 7.3.

One might contend that (3,4) hardly requires any special powers on SB's part, for P would obviously not try to prevent his best outcome from being implemented. But if this outcome is not, say, a Nash equilibrium, and there is another outcome that is or there is no Nash equilibrium (i.e., preferences are cyclical), then SB's powers may indeed be crucial in inducing (3,4).

In the large majority of these games (13 of the 16), there would seem to be no implementation problem: (3,4) is the unique Nash equilibrium and sometimes a nonmyopic equilibrium as well. These games might be considered *semi-undecidable*. The inferior outcome for SB is not as serious a problem as his 2 out-come in the undecidable games, but SB remains powerless to change it. Hence, an observer would not be able to say that this is not the outcome ordinary players would achieve, making SB indis-

---

2. In Section 4.4 I called such games—in which the outcome the stronger or quicker player can induce is ranked the same by both players—"fair," for they give him no special advantage; here, though, I now assume that superiority can be based on any of the three different kinds of power.

tinguishable as the (supposedly more powerful) player in such a game.

True, in the subset of these "(3,4) games" in which P does not have a dominant strategy, SB may be able to induce a better *initial outcome* for himself (i.e., 4) through his omniscience or deception. But even with his superior abilities, the sequential rules of play will enable P to move the process to (3,4) eventually, where it will remain.

Thus, despite his seemingly prodigious powers, SB does worse than P in 24 games, 8 of which are undecidable and 16 of which are semi-undecidable. In other words, in 31% of the 78 distinct 2×2 ordinal games, or 42% of the 57 games in which there is not a mutually best (4,4) outcome, SB's superiority does him no good—he cannot "catch up" to P if he happens to be the player in the disadvantageous position in these asymmetrical games.

It seems to me that there is no kind of power, short of giving SB the ability to *force* P's moves in some way, by which SB can rectify his position in these 24 games. This kind of forcing power, however, would violate P's presumed free will—the independent choices I have assumed he can make in a game—and vitiate a tenet of most western religions and existential philosophies that, at least in part, we are responsible for our own choices.

It is possible, of course, to suppose that SB is a veritable dictator and we are mere puppets. But if we cannot see the strings being pulled, SB is incontrovertibly undecidable. The presumption of being controlled, then, has no hope of ever being more than an article of faith, for we cannot prove the existence—or nonexistence—of something we cannot see, or perhaps even fathom.

In contradistinction to the puppeteer view, a SB who allows us some freedom—within the constraints of the superior (but nondictatorial) powers he enjoys over us—may open up opportunities for us to observe the exercise of this power, at least in the simple kinds of games I have postulated. (Observations need not be direct: we cannot literally see electrons and other elementary atomic particles, but physical theories provide a pattern into which these particles fit.) Yet a significant fraction of these games are undecidable or semi-undecidable. Another significant fraction, including the 21 with a mutually best (4,4) outcome, do not require any special exertions on SB's part.

This leaves only those games in which SB's M-power, S-power, or T-power induces an outcome he could not ensure without any of these qualities. Even these may not be telltale games, however, as I shall argue in Section 7.3.

7.3. *Arbitrariness and Undecidability*

In Section 7.2 I established that there are a number of games of total and partial conflict in which SB's combined powers, as I have defined them, would give him no leverage to improve the inferior outcome that would occur under the sequential rules of play. Three points of this argument may be challenged:

1. *Other definitions of power might provide SB with an escape.* This is possible, but, as I indicated previously, it would seem to require that SB directly control P's moves as well as his own. In addition to being philosophically unpalatable to me, this assumption, I think, leads to an intellectual dead end—decidability by fiat.

Instead, I prefer to investigate the consequences of supposing that P *can* make free and independent choices, restricted only by his awareness that his adversary in games of conflict, SB, may have certain superior powers. Checking for undecidability in this more liberal scheme of things is, I believe, an intellectually challenging and worthy task.

2. *SB may have no interest in improving an inferior outcome.* Again this is possible, but it would be incompatible with SB's having preferences in a game and acting rationally to achieve his best possible outcome. If SB is only a state of nature, without preferences, as assumed in the decision-theoretic analysis of Section 2.5, then "he" can never feel unhappy about not obtaining his best outcome, for he does not have one. In fact, he does not make choices—states simply occur, according to some chance mechanism, and they are presumably all the same to an indifferent nature.

But in the games I suppose SB plays with P, this is not the case. In the 24 undecidable and semi-undecidable games, in particular, SB obtains only his next-worst or next-best outcome, so he has reason to wish to do better.

Even if SB does not resent the fact that P does better than he does in these games, SB still falls short of obtaining his next-best outcome (in the undecidable games) and his best outcome (in the semi-undecidable games). If one builds resentment into SB's payoffs by assuming his inferior payoffs, vis-à-vis P's, diminish his status further,[3] he would be still more aggrieved, which would lower

3. Martin Shubik, Games of status, *Behavioral Sci.* 16, 2 (March 1971), 117–129.

his payoffs even more. Indeed, God expressed just such resentment to the prophet Samuel when He said, after the elders of Israel asked Samuel to "appoint a king for us, to govern us like all other nations" (1 Sam. 8:5): "It is not you that they have rejected; it is Me they have rejected as their king . . . forsaking Me and worshipping other gods" (1 Sam. 8:7-8).

I focused on the 24 undecidable and semi-undecidable games because of the resentment factor, which is not a problem in games like Prisoners' Dilemma in which the (3,3) compromise outcome is implemented.[4] In the (semi-)undecidable games, on the other hand, SB may have good reason to want to improve his ranking, because it is comparatively worse than P's.

Yet, if he cannot manage this feat with the powers I have assumed of him, then these games cannot be used as test cases to show whether SB is indeed a player. This is why I characterize them as (semi-)undecidable.

3. *Even granting that SB is a player with preferences and possesses the three kinds of power postulated, he may not play according to the sequential rules.* Rules I–IV and their variations, in my opinion, give the players maximum freedom, subject only to the restrictions that the strategies of *both* determine the outcome, and each selects his own strategy by himself. Remember that power accrues to SB because he can continue moving after P must stop (M-power), he can hold out longer and await P's initial strategy choice (S-power), or he can threaten to implement a Pareto-inferior outcome in repeated play (T-power) that will hurt both players in any single play. These powers seem not only plausible to me, but they also do not contravene, in any absolute way, P's freedom of movement.

Free choice, I think, implies sequentiality—the idea that one can always change one's mind and do something different, in response to what the other player does. I believe that the sequen-

---

4. Adding to these 24 games the seven fair games with (3,3) outcomes in which no kind of power is effective brings the total to 31, more than half of the 57 games of conflict without a mutually best (4,4) outcome. In other words, "powerlessness" on the part of SB is more the rule than the exception in the 2 × 2 ordinal games in which the players disagree about the best outcome. If one equates undecidability and powerlessness under the assumption that SB cannot be distinguished in such games—whether he ranks the (common) moving/staying/threat power outcome higher or lower or the same as P—SB's incognito status is greater than that suggested by the 24 (semi-)undecidable games discussed in the text. Also, if undecidability is defined in terms of powerlessness, it is rendered independent of interpersonal comparisons of rankings by the players, which are viewed as problematic by some theorists.

tial rules of play provide *the* most general framework within which to pose the issue of whether SB is undecidable. There may be other definitions of SB's powers that are compatible with essentially free movement by P, and I certainly do not want to preclude these. But my point here is that it is wise, I think, to start with rules of play that permit the players abundant flexibility of movement and then define SB's powers in terms of a tightening up of these rules. SB may encroach upon and even emasculate P's autonomy, but he is not a despot who can obliterate this autonomy altogether.

None of my previous arguments about SB's impotence in certain games, in spite of his superiority and presumed desire to do better, should be read as an argument for *universal* undecidability. On the contrary, many of the games discussed in this book, starting with the Revelation Game in Section 2.2 (Fig. 2.1), illustrate SB's efficacy, not inefficacy, in getting his way, in a manner of speaking. In the Revelation Game, for example, SB's M- and S-power enable him to implement his best outcome (4,2), which is not only better for him than the (2,3) Nash equilibrium but also better than the Pareto-superior (3,4) outcome, both of which favor P.

In Section 5.4 I suggested that sequential moves in the Revelation Game cry out for interpretation. Can SB reveal himself and then retract? Can P supplant nonbelief with belief? Are such moves and countermoves coordinated in play? Do they describe individual changes in religious belief, social movements of revival and decline, or perhaps only suggest that there is a fundamental instability underlying all belief in an ultimate reality?

I am inclined toward the latter view. If the Revelation Game at all parallels the preference rankings of the two players, stable *and* justified belief in SB is probably unattainable. Secure belief gives way to doubt, which plagues some individuals over their lifetimes and societies over the ages.

In the Revelation Game, recall, it is not rational for SB constantly to remain in plain view. By robbing P of an authentic choice, which I assume he always has in games and decisions, SB's choice of revelation precludes a true test of P's faith.

Admittedly, the Revelation Game and other games between SB and P may not occur in the limpid form I have presumed. The rational calculus may be a good deal more murky than is suggested by the crisp responses posited in the theory of moves.

This framework, nevertheless, seems to me to provide a succinct and fruitful perspective on the dynamic and continuing relationship between P and SB, whether SB is seen as a natural or supernatural figure. His possibly distinctive behavior in a game, and consequences this has for the outcome, determine his decidability.

Even in the decidable games, wherein SB can use his powers to induce a better outcome than an ordinary player can, it may be in his interest to incorporate an irreducible arbitrariness in his play. The Truth Game, discussed in Section 6.6 (Fig. 6.6), is a case in point. Like the Revelation Game, if SB is M- or S-powerful, he can induce (2,10), his best outcome and P's next-worst.

I assume that SB cannot dictate that P tell the truth. Using his M-power, however, SB may eventually coax it out of P. For after many cycles, P may give up trying doggedly to hide the truth, assuming that SB can invariably discern it and believe or not believe accordingly. Similarly, using S-power, SB can afford to await P's choice to ascertain whether he is being truthful; knowing he will be detected, P has an incentive not to lie.

But detection—certainly in the real world and perhaps in the supernatural world (as suggested by some biblical stories)—is often less than perfect. Furthermore, in the ethereal realm, one might ask whether (perfect) omniscience vitiates the free will many religions hold as a tenet: if our choices are truly free and independent, can God unerringly predict them without rendering them unfree? Earlier (Chapter 3, note 2) I suggested that this probably was not the case. However, if God's foreknowledge derives from His having worked things out as He wanted, which would seem necessary if His omniscience is not to stumble on human unpredictability (i.e., free choice), the answer is not so clear: there may indeed be a conflict between omniscience that makes humans predictable, on the one hand, and their having free will, on the other.

The essence of freedom, in my opinion, is not being captive of anybody—in thought or action. As I have shown, giving SB unsullied prescience can lead to the paradox of omniscience—at least in static (nonsequential) games—and its inferiority-producing results. The occasional perverse effects of S-power when P moves first in sequential games—or, equivalently, when SB predicts his strategy choice—highlights still another problem entailed by SB's omniscience.

If omniscience is untenable, partial omniscience—or, more simply, some modicum of predictability—certainly is not. I showed its desirable effects, for example, in the case of Prisoners' Dilemma when the players adhere to a choice rule of conditional cooperation. Such a tit-for-tat policy, however, creates problems for SB in the Truth Game, wherein, for the payoffs assumed, it is always rational for P to lie if SB strictly follows this policy, given that SB's detection equipment is not perfect.

SB can circumvent the lying problem by occasionally deviating from tit-for-tat. If he does so in the manner prescribed, he will not only raise his own expected payoff but also make it advantageous for P always to tell the truth.

SB can achieve this result in the strangest of ways—by paying only selective attention to the signal his detector gives him. In lowering his assumed reliability probability of 0.75 by responding to his detector according to certain derived probabilities, SB would appear to P to be acting arbitrarily, for there would be a random element in his response. Yet, this has the effect of inducing P to be truthful, which significantly helps SB and only slightly hurts P. In general, SB does better by paying only selective attention to his detector signal if his omniscience is less than complete.

In extending the notion of optimal mixed strategies from constant-sum games to variable-sum games with partial omniscience, SB's arbitrary behavior can be seen in a new light. First, it may be rational behavior in a broad class of situations in which his omniscience is not total.

Second, it can be used to induce strategy choices by P, thus differing from the classical concept of a mixed strategy that is designed solely to protect the player who uses it, independently of what the other player chooses. Indeed, the *raison d'être* of optimal mixed strategies in game theory is to minimize one's maximum losses (minimax), which simultaneously maximizes one's minimum gains (maximin) in constant-sum games.

By contrast, the "inducement" mixed strategy in the Truth Game guarantees SB a higher expected payoff than tit-for-tat only if P chooses his strategy of being truthful. However, SB can guarantee himself a somewhat less favorable expected payoff, whatever P does, by playing a minimax mixed strategy, based on his detection probability. Curiously, this mixed strategy, which creates an incentive for P to be truthful, benefits both players over what they would receive if SB resolutely and deterministically followed tit-for-tat.

Other unstable games, with neither Nash nor nonmyopic equilibria—or perhaps only one but not the other—may also offer their players mutual benefits when one player (presumably SB) acts arbitrarily, though in a way disciplined by his detection probability. The opportunities such games afford, of course, depend not only on certain mathematical conditions but also on whether these conditions have some reasonable interpretation in the play of a game.

I think the Truth Game can be so interpreted, whether the basis of SB's predictions are supernatural (in a theological context) or natural (in a secular context). The theological implications of arbitrary (random) behavior seem to me quite shattering, because heretofore the problem of evil and suffering in the world has been thought by some to undercut the existence of a benevolent God.

Of course, this depends on what one means by "benevolent." I take it to mean that God's purposes, if not virtuous, are certainly not reprehensible according to the normal canons of ethics. Although the Bible contains instances of evil (e.g., adultery, rape, and murder, including human sacrifice) that God seems to condone, one can certainly make the case that His behavior accomplishes some higher purpose to which we are not, and cannot be, privy.

To me this is not a wholly satisfactory argument, because it presupposes an ineradicable ignorance on our part and thereby precludes any conclusive test of whether God has a hand in earthly affairs. To put it another way, when we observe evil—contemptible as it is—God could well be acting in such a way as to minimize it (assuming it cannot be eliminated altogether), but how He does this we may never understand.

If this is a hard proposition to prove, it is given credence by the fact that SB's seemingly admirable purposes in the Truth Game—to discover the truth first, and then to favor truthful behavior on the part of P—are advanced by acting arbitrarily on occasion. One would not think that such guileful behavior could be used in the battle for truth, but it may help even the player (P) trying to hide it, depending on the mixed strategy SB adopts (inducement or minimax). In other words, diabolical means may cover up impeccable ends.

Arbitrariness, however, also accomplishes other more controversial purposes: (i) it makes SB appear more tentative, and (ii) it sanctions evil by making retribution—as well as rewards for good behavior—more uncertain. These by-products of a mixed strategy add to the mystery, which some people, of course, are delighted to see heightened (more on this below) but which may be hard to reconcile with a magnanimous God.

The accomplishment of (ii), particularly, makes the consequences of one's deeds and misdeeds less predictable. Without sure-fire rewards and punishment, risk-taking is encouraged, for one can afford to take more chances than is implied solely by SB's detection probability and a policy of tit-for-tat.

The accomplishment of (i) creates new undecidability problems for P. Is SB really ominscient/omnipotent/immortal, and, if not, to what degree does he possess these attributes? M- and S-power, as I showed, enable SB to implement his best outcome—when P is truthful and SB believes him—but it is not evident how, in practice, these kinds of power can be used in the Truth Game since it is one of imperfect information.

Partial omniscience, on the other hand, compounded by the uncertainty inherent in SB's optimal mixed strategy, makes the Truth Game appear considerably less decidable. In fact, I would argue, less-than-complete omniscience stalls the identification of SB in *any* game. This is true even in the subset of decidable games, like the Truth Game, wherein SB's superior qualities would appear to make a difference.

The reason is that SB's optimal strategy, being probabilistic, tends to camouflage his identity as a superior player in this game. Barring his wearing P down with M-power, or holding off a commitment with his S-power—both of which seem unrealistic choices in the Truth Game—SB's mixed strategy of both believing and not believing in what P says (and acting accordingly) will certainly appear perplexing, and perhaps malevolent, because it is not tightly bound to P's own actions.

Maybe this is as it should be—the divine mystery should be preserved, fostered, and perpetuated. If a clear pattern in SB's behavior emerged, then theological games would be flat and unengaging and, in the end, banal.

An aura of mystery magnifies their depth and fascination. That this aura can be supported by game-theoretic calculations that are optimal (in a particular sense) for the players, given their goals, is new, I believe, and perhaps a bit eerie. Significantly, one does not need to introduce uncertainty into undecidable and semi-undecidable games, because SB could not distinguish himself in these games anyway. Although he can do so in other games, like the Truth Game, it may pay for him (and contribute dividends to P as well) to inject some trace of the unknowable by acting arbitrarily.

Perhaps the most serious failing of unknowability and incomprehensibility is that it can be used to condone, if not justify, evil, making SB appear oblivious to suffering by not punishing the wicked for every sin they commit. This happens when the apparent risk of getting caught is not high enough to deter inimical behavior in any particular play of a game; on the other hand, SB may better satisfy estimable goals in the long run by his arbitrariness. But this obviates his sure-fire identification, even in decidable games, and may make him appear not just baffling but ruthless and morally corrupt as well.[5]

It is better, in my opinion, to try to understand the nature of the behavior that may underlie the Great Mystery than simply despair, or recoil in horror, at the problem of evil, incorrigible as it may seem. If the more mundane consequences of power and other asymmetries in our secular lives seem haphazard and hard to understand, is it any wonder that the perhaps supernatural powers exercised by SB might have deeply puzzling and troubling ramifications? Arbitrariness, however, as elusive and haunting as it is in our lives, may not be totally inexplicable.

I have not dispelled the Great Mystery, of course, for game theory is no *deus ex machina*. In a way, it becomes more impenetrable when recast in a game-theoretic framework that undergirds undecidability with a new kind of logic. The fact that games that are not already undecidable may be rendered so by the arbitrariness it may be rational for SB to introduce into their play obscures matters further. The enigma persists, though I hope, in this intellectual odyssey, that I have shed light on both its possible nature and why it remains shrouded.

To be sure, a world suffused with undecidability and arbitrariness may not be very comforting. Indeed, it may produce great personal distress. But I hope the discovery that there may be a logic to the apparent indeterminateness and incoherence of life and its choices may offer some relief.

Although I'm not sure that the lofty goals of a SB underlie life's vicissitudes, I believe such a possibility cannot be ruled out. This is perhaps a queer argument for agnosticism, for usually a sustained intellectual inquiry like this one reduces doubt and uncertainty. In this case it is heightened: theoretically, I have shown that there may be a pervasive rationality that supports undecidability and arbitrary behavior; empirically, such behavior seems unbiquitous. In short, both reason and evidence reinforce, I think, the Great Mystery. The intellectual challenge now is to probe and understand it further.

5. However, this is not to say that SB is intentionally remorseless or vindictive but rather that his behavior, insofar as it is judged to be amoral, is so because of his apparent desultory choices. Moral abandon, in other words, may be as much a product of caprice as cunning and legerdemain that are more transparent in their deviousness.

Appendix

There are 78 distinct 2 × 2 ordinal games in which the two players, each with two strategies, can strictly rank the four outcomes from best to worst. These games are "distinct" in the sense that no interchange of the column strategies, row strategies, players, or any combination of these can turn one game into another—that is, these games are structurally different with respect to these transformations.

Of the 78 games, 21 are no-conflict games with a mutually best (4,4) outcome. These outcomes are Nash and nonmyopic equilibria in these games, and no kind of power (moving, staying, or threat) would be needed by either player to implement them.

A listing of the remaining 57 games, in which the players disagree on a most-preferred outcome, is given below, with the numbers used in complete listings of the 78 games from two different sources given above each game.[1] I have also indicated, under the

---

1. Anatol Rapoport and Melvin Guyer, A taxonomy of 2 × 2 games, *General Systems: Yearbook of the Society for General Systems Research* 13 (1966), 195–201; and Steven J. Brams, Deception in 2 × 2 games, *J. Peace Sci.* 2, 2 (Spring 1977), 171–203. The first number above each game is that given in Brams; the second (in parentheses) is that given in Rapoport and Guyer.

most prominent games discussed in the text, the descriptive names used for these games: Chicken, Commitment, Prisoners' Dilemma, Punishment, Revelation, Testing, and Truth (ordinal ranking).

The 57 games are divided into two main categories: 26 (46%) in which at least one kind of power is effective; and 31 (54%) in which no kind of power is effective. Note that in all the games in which any of M-, S-, or T-power is effective, the player (column or row) who possesses this power can implement a better outcome for himself than when the other player possesses it, except for S-power in game 62(44).

The key to the symbols is as follows:

(x,y) = (row player, column player)
4 = best; 3 = next best; 2 = next worst; 1 = worst
Nash equilibria circled; * = nonmyopic equilibrium
U = undecidable game; SU = semi-undecidable game
m = M-power outcome for column; M = M-power outcome for row
s = S-power outcome for column; S = S-power outcome for row
t = T-power outcome for column; T = T-power outcome for row

---

26 Games in Which at Least One Kind of Power Is Effective

13(19)

(3,4)⃝ ^{m,s,t}	$(4,3)^{M,S,T}$
(1,2)	(2,1)

14(20)

(3,4)⃝ ^{m,s,t}	$(4,3)^{M,S,T}$
(2,2)	(1,1)

15(21)

(2,4)⃝ ^{m,s,t}	$(4,3)^{M,S,T}$
(1,2)	(3,1)

46(49)

(3,4)⃝ ^{m,s,t}	$(4,3)^{M,S,T}$
(2,1)	(1,2)

47(50)

(3,4)⃝ ^{m,s,t}	$(4,3)^{M,S,T}$
(1,1)	(2,2)

48(51)

(3,4)⃝ ^{m,s,t}	$(4,2)^{M,S}$
(2,1)	(1,3)

49(52)

(3,4)⃝ ^{m,s,t}	$(4,2)^{M,S}$
(1,1)	(2,3)

50(53)

(3,3)⃝ ^{m,s,t}	$(4,2)^{M,S}$
(2,1)	(1,4)

51(54)

(3,3)⃝ ^{m,s,t}	$(4,2)^{M,S}$
(1,1)	(2,4'

74(70)

$(3,4)^{m,s,t}$	$(2,1)$
$(4,2)^{M,S}$	$(1,3)$

75(71)

$(3,3)^{m,s,tT}$	$(2,1)$
$(4,2)^{M,S}$	$(1,4)$

66(56)

$(2,4)^{t}$	$(4,2)^{M,S}$
$(1,1)$	$(3,3)^{m,s,T}$

67(57)

$(2,3)$	$(4,2)^{M,S}$
$(1,1)$	$(3,4)^{m,s,t}$

Revelation

39(39)

$(2,4)^{*mM,sS,t}$	$(3,3)^{T}$
$(1,2)$	$(4,1)$

70(66)

$(3,3)^{*mM,sS}$	$(2,4)^{t}$
$(4,2)^{T}$	$(1,1)$

Chicken

58(73)

$(2,4)^{m,s}$	$(4,1)$
$(3,2)^{M,S}$	$(1,3)$

59(74)

$(2,4)^{m,s}$	$(3,1)$
$(4,2)^{M,S}$	$(1,3)$

60(75)

$(2,3)^{m,s}$	$(4,1)$
$(3,2)^{M,S}$	$(1,4)$

61(76)

$(2,3)^{m,s}$	$(3,1)$
$(4,2)^{M,S}$	$(1,4)$

Truth

65(55)

$(2,4)^{t}$	$(4,3)^{mM,sS,T}$
$(1,1)$	$(3,2)$

62(44)

$(2,4)^{S,t}$	$(4,1)$
$(1,2)$	$(3,3)^{mM,s,T}$

Punishment

68(64)

$(3,4)^{s,t}$	$(2,1)$
$(1,2)$	$(4,3)^{S,T}$

72(68)

$(2,2)$	$(3,4)^{sS,t}$
$(4,3)^{T}$	$(1,1)$

Commitment

73(69)

$(2,2)$	$(4,3)^{S,T}$
$(3,4)^{s,t}$	$(1,1)$

69(65)

$(2,4)^{t}$	$(3,1)$
$(1,2)$	$(4,3)^{sS,T}$

71(67)

$(2,3)$	$(3,4)^{s,t}$
$(4,2)^{S,T}$	$(1,1)$

31 Games in Which No Kind of Power is Effective

7(13) SU	**8(14) SU**	**9(15) SU**

(3,4) ^{mM,sS,t}	(4,2)	(3,4) ^{mM,sS,t}	(4,2)	(3,4) ^{mM,sS,t}	(4,1)
(2,3)	(1,1)	(1,3)	(2,1)	(2,3)	(1,2)

10(16) SU	**11(17) U**	**12(18) U**

(3,4) ^{mM,sS,t}	(4,1)	(2,4) ^{mM,sS,t}	(4,2)	(2,4) ^{mM,sS,t}	(4,1)
(1,3)	(2,2)	(1,3)	(3,1)	(1,3)	(3,2)

16(7)	**17(8)**	**18(9)**

(3,3) $^{*\,mM,sS}$	(4,2)	(3,3) $^{*\,mM,sS}$	(4,2)	(3,3) $^{*\,mM,sS}$	(4,1)
(2,4)	(1,1)	(1,4)	(2,1)	(1,4)	(2,2)

19(10) U	**20(11) U**	**21(12)**

(2,3) $^{*\,mM,sS}$	(4,2)	(2,3) $^{*\,mM,sS}$	(4,1)	(3,3) $^{*\,mM,sS,tT}$	(1,4)
(1,4)	(3,1)	(1,4)	(3,2)	(4,1)	(2,2)

Prisoners' Dilemma

31(31) SU	**32(32) SU**	**33(33) SU**

(3,4) $^{*\,mM,sS,t}$	(2,2)	(3,4) $^{*\,mM,sS,t}$	(2,1)	(3,4) $^{*\,mM,sS,t}$	(1,2)
(1,3)	(4,1)	(1,3)	(4,2)	(2,3)	(4,1)

34(34) SU	**35(35) U**	**36(36) U**

(3,4) ^{mM,sS,t}	(1,1)	(2,4) $^{*\,mM,sS,t}$	(3,2)	(2,4) $^{*\,mM,sS,t}$	(3,1)
(2,3)	(4,2)	(1,3)	(4,1)	(1,3)	(4,2)

37(37) SU	**38(38) SU**	**40(40) SU**

(3,4) $^{*\,mM,sS,t}$	(2,3)	(3,4) $^{*\,mM,sS,t}$	(1,3)	(3,4) ^{mM,sS,t}	(4,1)
(1,2)	(4,1)	(2,2)	(4,1)	(2,2)	(1,3)

41(41) SU

$(3,4)^{mM,sS,t}$	(4,1)
(1,2)	(2,3)

($(3,4)$ circled)

42(42)

$(3,3)^{mM,sS,t}$	(4,1)
(2,2)	(1,4)

($(3,3)$ circled)

43(43)

$(3,3)^{mM,sS,t}$	(4,1)
(1,2)	(2,4)

($(3,3)$ circled)

44(45) U

$(3,2)^{mM,sS}$	(4,1)
(2,3)	(1,4)

($(3,2)$ circled)

45(46) U

$(3,2)^{mM,sS}$	(4,1)
(1,3)	(2,4)

($(3,2)$ circled)

63(47) SU

(4,1)	(2,3)
$(3,4)^{mM,sS,t}$	(1,2)

($(2,3)$ circled)

68(48) SU

(4,1)	(2,2)
$(3,4)^{mM,sS,t}$	(1,3)

($(2,2)$ circled)

76(72) SU

(3,2)	(2,1)
$(4,3)^{mM,sS,T}$	(1,4)

77(77)

(2,2)	(4,1)
$(3,3)^{mM,sS,tT}$	(1,4)

Testing

78(78) SU

(2,2)	(3,1)
$(4,3)^{mM,sS,T}$	(1,4)

Glossary

This glossary contains definitions of game-theoretic and related terms used in this book. An attempt has been made to define these terms in relatively nontechnical language; more extended and rigorous definitions of some concepts (e.g., the three different kinds of power) can be found in the text.

Choice rule In a two-person game, a choice rule is a conditional strategy based on one player's prediction of the strategy choice of the other player.

Compellent threat In repeated play of a sequential two-person game, a threatener's compellent threat is to stay at a particular outcome to induce the threatenee to choose his (as well as the threatener's) best outcome associated with that strategy.

Complete information A game is one of complete information if the players know each others' preferences and the rules of play.

Conditional cooperation In a two-person game, conditional cooperation is a choice rule that says that a player will cooperate if he predicts the other player will cooperate; otherwise he will not.

Constant-sum (zero-sum) game A constant-sum game is a game in which the payoffs to the players at every outcome sum to some constant (or zero); all constant-sum games can be converted into zero-sum games by substracting the appropriate constant from the payoffs to the players.

Contingency A contingency is the set of strategy choices made by the player(s) other than the one in question.

Decision A decision is a game against nature (one-person game), in which rational players seek to make optimal choices based on different criteria.

Decision theory Decision theory is a mathematical theory for making optimal choices in situations in which the outcome does not depend on the choices of other players but rather on states of nature that arise by chance.

Deterrent threat In repeated play of a sequential two-person game, a threatener's deterrent threat is to move to another strategy to induce the threatenee to choose his (as well as the threatener's) best outcome associated with the threatener's initial strategy.

Dominant strategy A dominant strategy is a strategy that leads to outcomes at least as good as any other strategy for all possible contingencies, and a better outcome for at least one contingency.

Dominated strategy A dominated strategy is a strategy that leads to outcomes no better than those given by any other strategy for all possible contingencies, and a worse outcome for at least one contingency.

Effective power In a two-person game, a superior being's power is effective if he is able to induce a different (and usually better) outcome for himself when he possesses this power than when the other player possesses it.

Expected payoff/utility Expected payoff/utility is the sum of the payoff/utility a player receives from each outcome multiplied by its probability of occurrence, for all possible outcomes that may arise.

Fair game A fair game is a two-person game in which a superior being's power enables him to implement an outcome that both players rank the same.

Final outcome In a sequential game, the final outcome is the outcome induced by (possible) rational moves and countermoves from the initial outcome according to the theory of moves.

Game   A game is the sum-total of the rules of play that describe it.

Game against nature (one-person game)   A game against nature is a game in which one player is assumed to be "nature," whose choices are neither conscious nor based on rational calculation but on chance instead.

Game of partial conflict   A game of partial conflict is a variable-sum game in which the players' preferences are neither coincidental (game of total agreement) nor diametrically opposed (game of total conflict).

Game theory   Game theory is a mathematical theory of strategy to explicate optimal choices in interdependent decision situations, wherein the outcome depends on the choices of two or more actors, or players.

Game of total agreement   A game of total agreement is a variable-sum game in which the preferences of the players for all outcomes coincide.

Game of total conflict   A game of total conflict is a constant-sum game in which what one player gains the other player(s) lose.

Game tree   A game tree is a symbolic tree based on the rules of play of the game, in which the vertices of the tree represent choice points and the branches represent alternative courses of action that can be chosen.

Incomprehensibility   A player renders himself incomprehensible when he uses a mixed (or random) strategy that makes his behavior appear arbitrary.

Inducement strategy   In a two-person game in which one player has partial omniscience, an inducement strategy is one in which an inducer offers an incentive to an inducee to choose a particular strategy, in order to maximize his expected payoff/utility, that also favors the inducer.

Initial outcome   In a sequential game, the initial outcome is the outcome rational players choose when they make their initial strategy choices according to the theory of moves.

Lexicographic decision rule   A lexicographic decision rule enables a player to rank outcomes on the basis of a most important criterion ("primary goal"), then a next most important criterion ("secondary goal"), and so forth.

Minimax strategy   In a two-person game, a minimax strategy ensures a player of the value in a constant-sum game; in a variable-sum game in which one player has partial omniscience, it guarantees this player a particular minimum expected payoff/ utility independent of the strategy choice of the other player.

Mixed strategy In a normal-form game, a mixed strategy is a strategy that involves the random selection from two or more pure strategies, according to a particular probability distribution.

Moving power (omnipotence) In a two-person sequential game, moving power is the ability to continue moving when the other player must eventually stop.

Nash equilibrium In a normal-form game, a Nash equilibrium is an outcome from which no player would have an incentive to depart unilaterally because he would do (immediately) worse, or at least not better, if he moved.

Newcomb's problem Newcomb's problem describes an apparent conflict between two principles of choice: dominance and expected utility.

Nonmyopic calculation In a two-person sequential game, nonmyopic calculation assumes that rational players make choices in full anticipation of how each will respond to the other, both in selecting their strategies initially and making subsequent moves.

Nonmyopic equilibrium In a two-person sequential game, a nonmyopic equilibrium is an outcome from which neither player, anticipating all possible rational moves and countermoves from the initial outcome, would have an incentive to depart because he would do (eventually) worse, or at least not better, if he did.

Normal form A game in normal form is represented by an outcome/payoff matrix in which players are assumed to choose their strategies independently.

Omniscience Omniscience is the ability of a player in a two-person game to predict the other player's strategy choice (before he makes it); omniscience is partial when the prediction is correct with probability less than one.

Ordinal game An ordinal game is a game in which the players can rank, but not necessarily assign payoffs or utilities, to the outcomes.

Outcome/payoff matrix In a game in normal form, an outcome/payoff matrix is a rectangular array, or matrix, whose entries indicate the outcomes/payoffs to each player resulting from each of their possible strategy choices.

Paradox of inducement   In a two-person game in which one player has a dominant strategy and the other player does not, a paradox of inducement occurs when the player without a dominant strategy is able to induce an outcome he ranks higher than the player with a dominant strategy.

Paradoxes of omniscience   In a two-person game in which one player can predict the strategy choice of the other player and the other player is aware of the first player's omniscience, a paradox of omniscience occurs if (1) the player without omniscience is able to obtain his best outcome but the player with omniscience is not, or (2) both players obtain a Pareto-inferior outcome.

Paradox of (staying) power   In a two-person sequential game, a paradox of staying power occurs when the superior being who possesses this power does worse than when the other player possesses it.

Pareto-inferior/superior outcome   An outcome is Pareto-inferior if there exists another outcome that is better for some player(s) and not worse for all the other player(s). If there is no such other outcome, the outcome in question is Pareto-superior.

Payoff   See Utility.

Perfect information   A game is one of perfect information if each player knows with certainty the strategy choice or move of every other player at each point in the sequence of play.

Preference   The preference of a player is his ranking of outcomes from best to worst.

Pure strategy   In a game in normal form, a pure strategy is a single strategy.

Rational player   A rational player is one who seeks to attain better outcomes, according to his preference, in light of the presumed rational choices of other players in a game or of the states of nature that may arise in a decision.

Rational termination   Rational termination is a constraint, assumed in the definition of staying power, that prohibits the player without such power from moving from the initial outcome if it leads to cycling back to this outcome in a sequential game.

Revealed deception   In a game of incomplete information, revealed deception involves a deceiver's falsely announcing one strategy to be dominant, but subsequently choosing another strategy, thereby revealing his deception.

Rules of play The rules of play of a game describe the preferences and choices available to the players, their sequencing, and any special prerogatives one player (such as a superior being) may have in the play of the game.

Security level In a normal-form game, the security level of a player is the best outcome or payoff he can ensure for himself, whatever strategies the other players choose.

Sequential game A sequential game is one in which players can move and countermove after their initial strategy choices according to the theory of moves.

State of nature A state of nature is a situation or set of circumstances that arises by chance in the world.

Staying power (omnipotence) In a two-person sequential game, staying power is the ability of a player to hold off making a strategy choice until the other player has made his.

Strategy In a game in normal form, a strategy is a complete plan that specifies all possible courses of action of a player for whatever contingencies may arise.

Supergame A supergame is a game that comprises repeated plays of a nonsequential game.

Superior being In a two-person game, a superior being is a player who possesses one or more of the attributes of omniscience, omnipotence (moving power and staying power), immortality (threat power), and incomprehensibility.

Symmetrical game A symmetrical game is a two-person game in which the ranks of the outcomes by the players along a diagonal are the same, whereas the ranks of the off-diagonal outcomes are mirror images of each other.

Tacit deception In a game of incomplete information, tacit deception involves a deceiver's falsely announcing one strategy to be dominant, and subsequently choosing this strategy, so as not to reveal his deception.

Theory of moves The theory of moves describes optimal strategic calculations in normal-form games in which the players can move and countermove from an initial outcome in sequential play.

Threat power (immortality) In a two-person sequential game that is repeated, threat power is the ability of a player to threaten a mutually disadvantageous outcome in the single play of the game to deter untoward actions in the future play of this or other games.

Undecidability Undecidability characterizes two-person games in which an ordinary being cannot decide whether the other player is superior from the outcome that is induced. Undecidable games are games in which the superior being obtains his next-worst outcome, whereas semi-undecidable games are games in which he obtains his next-best outcome, with the ordinary being in both cases obtaining an outcome that he ranks higher than the superior being.

Undominated strategy An undominated strategy is a strategy that is neither unconditionally best, or dominant, nor unconditionally worst, or dominated.

Utility Utility is the numerical value, indicating degree of preference, that a player attaches to an outcome in a decision or game.

Value In a two-person game, the value is the payoff that a player can ensure for himself whatever the other player does; in constant-sum games, this value is a common one (except for a reversal in sign) guaranteed by a minimax strategy, whereas in variable-sum games the players will in general have different values, or security levels.

Variable-sum game A variable-sum game is a game in which the sum of the payoffs to the players at different outcomes varies, so the players may gain or lose simultaneously at different outcomes.

Zero-sum game See Constant-sum (zero-sum) game.

Bibliography

Anglin, W. S., Can God create a being he cannot control?, *Analysis* 40, 4 (October 1980), 220–223.

Augustinus, Aurelius, Saint, *The Confessions of Saint Augustine*, translated by Edward B. Pusey (New York: Modern Library, 1949).

Bachrach, Peter, and Morton S. Baratz, Decisions and nondecisions: an analytical framework, *Am. Political Sci. Rev.* 57, 3 (September 1963), 632–642.

Bachrach, Peter, and Morton S. Baratz, Two faces of power, *Am. Political Sci. Rev.* 56, 4 (December 1962), 947–952.

Barth, Karl, *Church Dogmatics*, translated by G. W. Bromily (New York: Harper & Row, 1962).

Brams, Steven J., Belief in God: a game-theoretic paradox, *Int. J. Philos. Religion* 13, 3 (1982), 121–129.

Brams, Steven J., *Biblical Games: A Strategic Analysis of Stories in the Old Testament* (Cambridge, MA: MIT Press, 1980).

Brams, Steven J., Deception in 2 × 2 games, *J. Peace Sci.* 2 (Spring 1977), 177–203.

Brams, Steven J., *Game Theory and Politics* (New York: Free Press, 1975).

Brams, Steven J., Mathematics and theology: game-theoretic implications of God's omniscience, *Math. Mag.* 53, 5 (November 1980), 277–282; comment by Ian Richards, *Math. Mag.* 54, 3 (May 1981), 148; and reply by Brams, *Math. Mag.* 54, 4 (September 1981), 219.

Brams, Steven J., Newcomb's Problem and Prisoners' Dilemma, *J. Conflict Resolution* 19, 4 (December 1975), 596–612.

Brams, Steven J., Omniscience and omnipotence: how they may help—or hurt—in a game, *Inquiry* 25, 2 (June 1982), 217–231.

Brams, Steven J., *Paradoxes in Politics: An Introduction to the Nonobvious in Political Science* (New York: Free Press, 1976).

Brams, Steven J., A resolution of the paradox of omniscience. In *Reason and Decision*, Bowling Green Studies in Applied Philosophy, Vol. III-1981, ed. Michael Bradie and Kenneth Sayre (Bowling Green, OH: Department of Philosophy, Bowling Green State University, 1982), pp. 17–30.

Brams, Steven J., Morton D. Davis, and Philip D. Straffin, Jr., The geometry of the arms race, *Int. Studies Quarterly* 23, 4 (December 1979), 567–588; Raymond Dacey, Detection and disarmament, pp. 589–598; Brams, Davis, and Straffin, A reply to "Detection and disarmament," pp. 599–600.

Brams, Steven J., and Marek P. Hessel, Absorbing outcomes in 2 × 2 games, *Behavioral Sci.* 27, 4 (October 1982), 393–401.

Brams, Steven J., and Marek P. Hessel, Staying power in sequential games, *Theory and Decision* (forthcoming).

Brams, Steven J., and Marek P. Hessel, Threat power in sequential games (mimeographed, 1982).

Brams, Steven J., and Donald Wittman, Nonmyopic equilibria in 2 × 2 games, *Conflict Management Peace Sci.* 6, 1 (1983).

Brams, Steven J., and Frank C. Zagare. Deception in simple voting games, *Social Sci. Res.* 6, 3 (September 1977), 257–272.

Brams, Steven J., and Frank C. Zagare, Double deception: two against one in three-person games, *Theory and Decision* 13, 1 (March 1981), 81–90.

Buber, Martin. *I and Thou*, 2nd ed., translated by Ronald Gregor Smith (New York: Scribner's, 1958).

Contemporary Philosophy of Religion, ed. Stephen M. Cahn and David Shatz (New York: Oxford University Press, 1982).

Dacey, Raymond, Detection, inference and the arms race. In *Reason and Decision*, Bowling Green Studies in Applied Philosphy, Vol. III-1981, ed. Michael Bradie and Kenneth Sayre (Bowling Green, OH: Department of Philosophy, Bowling Green State University, 1982), pp. 87–100.

Davis, Philip J., and Reuben Hersh, *The Mathematical Experience* (Boston: Birkhäuser, 1980).

DeSua, Frank, Consistency and completeness—a resumé, *Am. Math. Monthly* 56, 5 (May 1956), 295–305.

Eells, Ellery, *Rational Decision and Causality* (Cambridge: Cambridge University Press, 1982).

Einstein, Albert, *Out of My Later Years* (New York: Philosophical Library, 1950).

Engelbrecht-Wiggans, Richard, and Robert J. Weber, Notes on a sequential auction involving asymmetrically-informed bidders, *Int. J. Game Theory* (forthcoming).

Eves, Howard, *Great Moments in Mathematics (after 1650)* (Washington, DC: Mathematical Association of America, 1981).

Fackenheim, Emil L., An outline of modern Jewish theology. In *Faith and Reason: Essays in Judaism*, ed. Robert Gordis and Ruth B. Waxman (New York: KTAV, 1973), pp. 211–220.

Ferejohn, John A., Personal communication, May 27, 1975.

Fishburn, Peter C., Lexicographic orders, utilities and decision rules: a survey, *Management Sci.* 20, 11 (July 1974), 1442–1471.

Frank, Phillip, *Einstein: His Life and Times* (New York: Knopf, 1947).

Fraser, Niall M., and Keith W. Hipel, Solving complex conflicts, *IEEE Trans. Systems, Man, Cybernetics* SCM-9, 12 (December 1979), 805–816.

Gardner, Martin, *The Ambidextrous Universe: Mirror Asymmetry and Time-Reversed Worlds*, 2nd ed. rev. (New York: Scribner's, 1979).

Gardner, Martin, Mathematical games, *Scientific American* (July 1973), 104–108.

Gardner, Martin (written by Robert Nozick), Mathematical games, *Scientific American* (March 1974), 102–108.

Goldstine, Herman, *The Computer from Pascal to von Neumann* (Princeton, NJ: Princeton University Press, 1972).

Hanson, Norwood Russell, *What I Don't Believe and Other Essays*, ed. Stephen Toulmin and Harry Woolf (Dordrecht, Holland: D. Reidel, 1971).

Henderson, John M., and Richard E. Quandt, *Microeconomic Theory: A Mathematical Approach*, 2nd ed. (New York: McGraw-Hill, 1971).

Ho, Y. C., A simple example on informativeness and competitiveness, *J. Optimization Theory Appl.* 11, 4 (April 1973), 437–440.

Hofstadter, Douglas R., *Gödel, Escher, Bach: An Eternal Golden Braid* (New York: Basic, 1979).

Howard, Nigel, *Paradoxes of Rationality: Theory of Metagames and Political Behavior* (Cambridge, MA: MIT Press, 1971).

Howard, Nigel, Personal communications, March 27, 1975, and June 25, 1975.

Jaynes, Julian, *The Origin of Consciousness in the Breakdown of the Bicameral Mind* (Boston: Houghton Mifflin, 1976).

Jewish Publication Society, *The Prophets* (Philadelphia: Jewish Publication Society, 1978).

Jewish Publication Society, *The Torah: The Five Books of Moses*, 2nd ed. (Philadelphia: Jewish Publication Society, 1967).

Jewish Publication Society, *The Writings* (Philadelphia: Jewish Publication Society, 1982).

Jones, J. P., Some undecidable determined games, *Int. J. Game Theory* 11, 2 (1982), 63–70.

Kac, Mark, and Stanislaw M. Ulam, *Mathematics and Logic: Retrospect and Prospects* (New York: New American Library, 1969).

Kaplan, Abraham, The Jewish argument with God, *Commentary* 70 (October 1980), 43–47.

Kenney, Anthony, *The God of the Philosophers* (Oxford: Clarendon, 1979).

Kilgour, D. Marc, Equilibria for far-sighted players, *Theory and Decision* (forthcoming).

Kolakowski, Leszek, *Religion* (New York: Oxford University Press, 1982).

Kolata, Gina, Does Gödel's Theorem matter to mathematics?, *Science* 218, 4574 (19 November 1982), 779–780.

Kramer, Edna E., *The Nature and Growth of Modern Mathematics*, Vol. 2 (Greenwich, CT: Fawcett, 1970).

Küng, Hans, *Does God Exist? An Answer for Today*, translated by Edward Quinn (New York: Doubleday, 1980).

Küng, Hans, *Freud and the Problem of God*, translated by Edward Quinn (New Haven, CT: Yale University Press, 1979).

Küng, Hans, *Justification: The Doctrine of Karl Barth and a Catholic Reflection* (Philadelphia: Westminister, 1981).

Kushner, Harold S., *When Bad Things Happen to Good People* (New York: Schocken, 1981).

Lefebvre, Vladimir A., *Algebra of Conscience: A Comparative Analysis of Western and Soviet Ethical Systems* (Dordrecht, Holland: D. Reidel, 1982).

Lefebvre, Vladimir A., Personal communication, March 25, 1980.

Lewis, David, Prisoners' Dilemma is a Newcomb problem, *Philos. Public Affairs* 8, 3 (Spring 1979), 235–240.

Levi, Isaac, A note on Newcombmania, *J. Philos.* 79, 6 (June 1982), 337–342.

The Logic of God: Theology and Verification, ed. Malcolm L. Diamond and Thomas F. Litzenburg, Jr. (Indianapolis: Bobbs-Merrill, 1975).

Mavrodes, George I., Rationality and religious belief—a perverse question. In *Rationality and Religious Belief*, ed. C. F. Delaney (Notre Dame, IN: University of Notre Dame Press, 1979), pp. 28–41.

The Mind's I: Fantasies and Reflections on Self and Soul, ed. Douglas R. Hofstadter and Daniel C. Dennett (New York: Basic, 1981).

Mind Design: Philosophy, Psychology, Artificial Intelligence, ed. John Haugeland (Cambridge, MA: MIT Press, 1981).

Muzzio, Douglas, *Watergate Games: Strategies, Choices, Outcomes* (New York: New York University Press, 1982).

Nagel, Ernest, and James R. Newman, *Gödel's Proof* (New York: New York University Press, 1968).

Nash, John, Non-cooperative games, *Ann. Math.* 54 (1951), 286–295.

Nozick, Robert, Newcomb's Problem and two principles of choice. In *Essays in Honor of Carl G. Hempel*, ed. Nicholas Rescher (Dordrecht, Holland: D. Reidel, 1969), pp. 114–146.

Pascal's Pensées, translated by H. F. Stewart (New York: Pantheon, 1950).

Political Power: A Reader in Theory and Research, ed. Roderick Bell, David V. Edwards, and R. Harrison Wagner (New York: Free Press, 1969).

Popper, Karl R., *The Open Universe: An Argument for Indeterminism*, ed. W. W. Bartley, III (Totowa, NJ: Roman and Littlefield, 1982).

The Power of God: Readings on Omnipotence and Evil, ed. Linwood Urban and Douglas N. Walter (New York: Oxford University Press, 1978).

Rapaport, Anatol, and Albert Chammah, *Prisoner's Dilemma: A Study in*

Conflict and Cooperation (Ann Arbor, MI: University of Michigan Press, 1965).

Rapoport, Anatol, and Melvin Guyer, A taxonomy of 2 × 2 games, *General Systems: Yearbook of the Society for General Systems Research* 11 (1966), 203–214.

Rationality and Religious Belief, ed. C. F. Delaney (Notre Dame, IN: University of Notre Dame Press, 1979).

Rosenkrantz, Gary, and Joshua Hoffman, What an omnipotent agent can do, *Int. J. Philos. Religion* 11, 1 (Spring 1980), 1–19.

Roy, Rustom, *Experimenting with Truth: The Fusion of Religion with Technology, Needed for Humanity's Survival* (Oxford: Pergamon, 1981).

Rubinstein, Ariel, A note on the duty of disclosure, *Economic Lett.* 4 (1979), 7–11.

Rucker, Rudy, *Infinity and the Mind: The Science and Philosophy of the Infinite* (Boston: Birkhäuser, 1982).

Schelling, Thomas C., *Arms and Influence* (New Haven, CT: Yale University Press, 1966).

Shubik, Martin, Games of status, *Behavioral Sci.* 16, 2 (March 1971), 117–129.

Shubik, Martin, *Game Theory in the Social Sciences: Concepts and Solutions* (Cambridge, MA: MIT Press, 1982).

Some Strangeness in the Proportion: A Centennial Symposium to Celebrate the Achievements of Albert Einstein, ed. Harry Woolf (Reading, MA: Addison-Wesley, 1980).

Sontag, Frederick, *The God of Evil: An Argument from the Existence of the Devil* (New York: Harper & Row, 1970).

Spinoza, Benedictus de, *Chief Works,* translated by R. H. M. Elwes (New York: Dover, 1951).

Swinburne, Richard, *The Existence of God* (Oxford: Clarendon, 1979).

Swinburne, Richard, *Faith and Reason* (Oxford: Clarendon, 1981).

Turing, A. M., Computing machinery and human intelligence, *Mind* 59, 236 (October 1950), 433–460.

von Neumann, John, and Oskar Morgenstern, *Theory of Games and Economic Behavior,* 3rd ed. (Princeton, NJ: Princeton University Press, 1953).

Wierenger, Edward. Omnipotence defined, *Philos. and Phenomenological Res.* 43 (1982) (forthcoming).

Zagare, Frank C., A game-theoretic analysis of the Vietnam negotiations: preferences and strategies 1968–1973, *J. Conflict Resolution* 21, 4 (December 1977), 663–684.

Zagare, Frank C., The Geneva Conference of 1954: a case of tacit deception, *Int. Studies Quarterly* 23, 3 (September 1979), 390–411.

Index